Adenomatous Polyps of the Colon

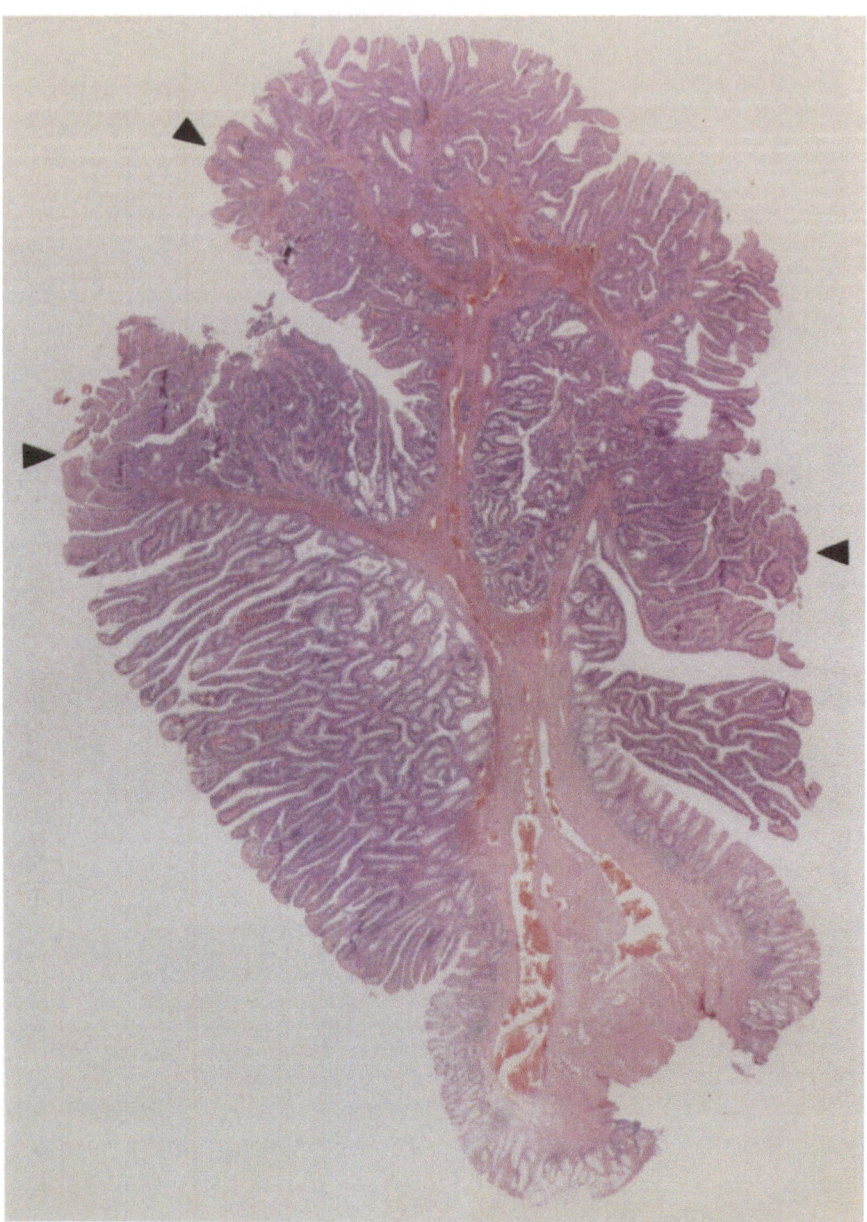

Malignant polyp arising in tubulovillous adenoma. There is a large area of carcinoma between the three arrowheads. Above and to the left of the lowest arrowhead the carcinoma has broken through the muscularis mucosae (the thin vertical red line) into the submucosa of the head of the polyp. The stalk and cauterized margin at the bottom are free of tumor. H&E; × 11 reproduced at 69%.

Robert Lev

Adenomatous Polyps of the Colon

Pathobiological and Clinical Features

With Contributions by M. Peter Lance

With a Foreword by B.C. Morson

With 35 Illustrations, 5 in Full Color

Springer Science+Business Media, LLC

ROBERT LEV, M.D., Professor of Pathology, Brown University Program in Medicine; Associate Pathologist, Roger Williams General Hospital, Providence, Rhode Island 02908, USA

M. PETER LANCE, M.A., M.B., M.R.C.P., Assistant Professor of Medicine, State University of New York at Buffalo; Chief, GI Section, Veterans Administration Medical Center, Buffalo, New York 14215, USA

Library of Congress Cataloging in Publication Data
Lev, Robert.
   Adenomatous polyps of the colon : pathobiological and clinical
features / Robert Lev, with a contribution by M. Peter Lance ; with
a foreword by B.C. Morson.
     p.   cm.
   Includes bibliographical references.

   1. Intestinal polyps—Histopathology. 2. Intestinal polyps—
—pathophysiology.   I. Lance, M. Peter   II. Title.
   [DNLM: 1. Adenoma. 2. Colonic Neoplasms. 3. Colonic Polyps.   WI
520 L655a]
RC 280.I5L48   1990
616.99'3347—dc20
DNLM/DLC
for Library of Congress   89-21869
                                                CIP

Typeset by Caliber Design Planning, Inc., New York, New York.

9 8 7 6 5 4 3 2 1

ISBN 978-1-4612-6453-8   ISBN 978-1-4419-8706-8 (eBook)
DOI 10.1007/978-1-4419-8706-8

To my parents,
Sarah and Joseph Lev,
and
to Ori Sherman

Foreword

The disciplines of epidemiology and genetics, which make such an important contribution to our knowledge of colorectal cancer, cannot operate effectively without agreement on the histopathological classification of colonic polyps and their malignant potential. Colorectal adenoma is the histopathological type of polyp with significant potential for malignant change; therefore, management of differing magnitudes of risk for colorectal cancer in various categories of patients with adenomas is essential if we are to design cancer prevention programs that are clinically relevant and cost-effective. From overwhelming evidence it has been shown that the adenoma is the principal precursor lesion for colorectal cancer and that removal of adenomas can be equated with the prevention of this common malignancy. Putting this theoretical assumption into practice requires detailed knowledge of the pathology and natural history of the adenoma–carcinoma sequence as well as those clinical circumstances known to predispose patients to colorectal cancer, such as adenomatous polyposis, a positive family history of colorectal carcinoma, and the metachronous carcinoma problem. The establishment of registers of patients with precancerous conditions, such as the National Polyp Study and the Polyposis Registries, is an important development because they not only serve as a store of information and a focus for research but can be responsible for the design and coordination of screening programs. All these issues are addressed in this volume, which is a most welcome addition to the literature of cancer prevention and the prevention of colorectal cancer in particular.

B.C. Morson, C.B.E, V.R.D, M.A., D.M., F.R.C.Path, F.R.C.S., F.R.C.P.

Preface

It is of the highest importance . . .
to be able to recognize out
of a number of facts which are
incidental and which vital.

Sherlock Holmes
(A. Conan Doyle)

The main purpose of this effort is to consolidate into one volume the vast amount of information that has accumulated on the basic science and clinical aspects of adenomatous polyps (APs) of the colon. Numerous articles, monographs, and symposia have appeared on the subject of APs in the past decade, many of which have been inspired by the increasing realization that most colorectal carcinomas appear to arise in such polyps. However, the information is scattered throughout the literature, and few attempts have been made to integrate the scientific and clinical components. We will try to evaluate these data, with special emphasis on their implications for the management of subjects bearing such polyps.

Dr. Peter Lance, the contributor to this volume, is a gastroenterologist who has had extensive experience with sigmoidoscopy and colonoscopy and who has performed a number of research studies on APs and ulcerative colitis. He has provided the clinical input into this endeavor. Since interpretation of risk factors and adequacy of polyp excision are critical to future therapy and surveillance for polyp-bearers, the timing and technique of polypectomy are important issues. He will describe and evaluate the various endoscopic procedures and briefly discuss the costs involved in screening and surveillance.

Dr. Alan Morrison, Professor of Medical Science in the Department of Community Health, Brown University, has provided material for and critical commentary on the sections on natural history of polyps, epidemiology, and screening. He has also contributed Figure 4.2, which deals with the adenoma–carcinoma sequence, to the section on etiology of polyps.

My task will be to summarize the pathological and biological characteristics of APs. I will emphasize those features of polyps that may affect therapy. I will also discuss, with illustrations, certain problems in the classification of polyps pertaining to risk factors that may affect management of both the polyp and the patient. These include grading schemes for the degree of dysplasia and villosity in APs. Differentiation of adenomatous from regenerative epithelium as well as distinctions between adenomatous and hyperplastic polyps will also receive considerable attention.

We hope that this enterprise will prove to be informative and useful to the reader.

Robert Lev, M.D.

Acknowledgments

Dr. Basil Morson, who graciously wrote the Foreword to this book, also provided useful comments on malignant polyps. I would like to thank Dr. Daniel Wroblewski, Section of Colorectal Surgery, and Drs. Herbert Rakatansky and Thomas DeNucci, Section of Gastroenterology, Roger Williams General Hospital, for helpful discussions of various aspects of polyp identification and management. Dr. Wroblewski also kindly provided specimens for some of the illustrative material used in this book, as have numerous colleagues over the past 20 years. The electron micrograph of adenomatous epithelium (Figure 2.23) was obtained courtesy of Dr. Shirley Siew. Discussions and slide review with Dr. John Abbott were helpful in formulating criteria for grading dysplasia in polyps.

Current results from the St. Marks adenoma follow-up study and from the Erlangen Polyp Registry were kindly furnished by Dr. C.B. Williams and Dr. P. Hermanek, respectively. The data on the Rhode Island Polyp Registry were obtained with the assistance of Susan Rehkopf, Tumor Registrar, and Lorna Gabrielli, Assistant Registrar.

Linda Mulzer provided invaluable assistance with the preparation of the manuscript, as did Paul Camara with the graphics and Marion Poisson, Audrey Gagne, and David Niedel-Gresh with the photographs. The part of the literature review involving Medlars II (National Library of Medicine Retrieval Service) was carried out by Hadassah Stein, Health Services Librarian at this hospital.

Contents

Introduction

In reviewing the vast literature on colonic adenomatous polyps (APs) several problems become apparent. First, there are major differences between data obtained from autopsy and data obtained from clinical studies, some of which reflect the older age of the autopsied population. For example, the autopsy studies reveal a higher prevalence of polyps, a higher percentage of smaller lesions, and a more even distribution throughout the colon than is found in clinical (surgical and endoscopic) series. Polyps in segments of colon resected because of large or histologically suspicious polyps, or those found incidentally during colectomies for carcinoma, are larger than those noted in endoscopic studies of asymptomatic subjects. The endoscopic group can in turn be divided into precolonoscopic and colonoscopic (from the early 1970s onward) periods. The precolonscopic studies, using primarily the rigid sigmoidoscope with its range of under 25 cm of distal large intestine (often < 20 cm in practice, Nivatvongs & Fryd, 1980), provided data only on the rectum and a portion of the sigmoid colon. Another problem is that of variable definition of the anatomic segments of the colon. Length of rectum and sigmoid colon, site of the junction between the two, and what constitutes the "left" colon (e.g., pre- or postsplenic flexure) are not always clearly described and often make it difficult to compare the results of different authors. Other potentially confounding factors include use of flexible sigmoidoscopes in some studies and rigid instruments in others and the persistent use of the single contrast barium enema by some radiologists rather than the currently favored double contrast enema. Finally, many earlier studies failed to differentiate between hyperplastic polyps, believed by the majority of investigators to have no malignant potential, and adenomatous polyps.

Adenomatous polyps are significant mainly because (1) they may undergo malignant transformation and (2) persons bearing them are prone to develop metachronous (future) colorectal carcinomas (CRCs). The pathological and epidemiological evidence for this malignant potential will therefore be discussed in great detail. Special attention will be paid to risk factors (e.g., multiplicity, large size, severe dysplasia, and significant villous component) that help determine prognosis and surveillance modalities in subjects with polyps. It has become apparent that some of these risk factors not only are associated with cancer in that polyp but are also predictive of metachronous CRCs in those individuals.

Considerable space will also be devoted to other possible precursors of CRC. One of the best characterized of these is the severe mucosal epithelial dysplasia found in ulcerative colitis (UC). In fact, there is a histological similarity between the severe dysplasia that occurs in UC and in APs, suggesting to B.C. Morson (Konishi & Morson, 1982) and others that the dysplasia–carcinoma sequence is applicable in both settings. This terminology might reduce the controversy embodied in the term *polyp–cancer* (or *adenoma–cancer*) *sequence*. Possible de novo carcinomas arising in flat mucosa, an event responsible for an unknown number of CRCs, and other conditions such as hereditary nonpolyposis colorectal cancer, now believed to account for up to 5 percent of CRCs, will also be discussed at length. The latter condition is of great current interest because of advances in genetics and recent detailed genealogical studies that indicate that familial aggregates of CRCs and APs are more common than previously suspected.

Despite these controversies and problems in interpreting the literature, I hope in the following chapters to weave a coherent story about colonic polyps and their implications. The thesis I would like to develop is that since most CRCs arise in adenomatous polyps, such polyps, especially those associated with the risk factors mentioned above, present us with the most common and significant biological marker for metachronous colonic carcinoma that currently exists. It would appear to follow from this that detection and eradication of such polyps, followed by extended, perhaps lifelong, surveillance of patients bearing them, should be the most practical way of reducing incidence of and mortality from CRC. Unfortunately, there are two major problems with this proposal. First, and more important, is the inability of screening programs thus far to produce this reduced mortality from CRC. Second, the financial and manpower costs of detecting and removing polyps are considerable. These points will be discussed extensively in the text.

References

Konishi F, Morson B (1982) Pathology of colorectal adenomas: A colonoscopic survey. J Clin Pathol 35:830–841.
Nivatvongs S, Fryd D (1980) How far does the proctosigmoidoscope reach? A prospective study of 1000 patients. N Engl J Med 303:380–382.

Color Plates for Chapter 2

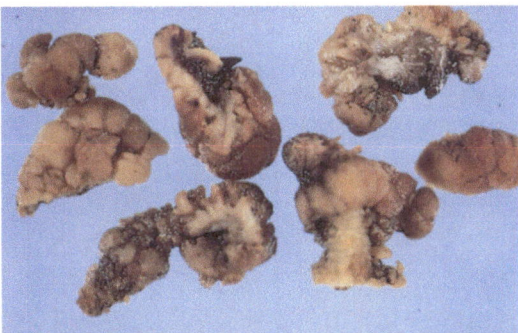

FIGURE 2.1. Multiple tubular adenomas showing smooth, lobulated surfaces. Multiplicity is a risk factor for synchronous and metachronous carcinomas.

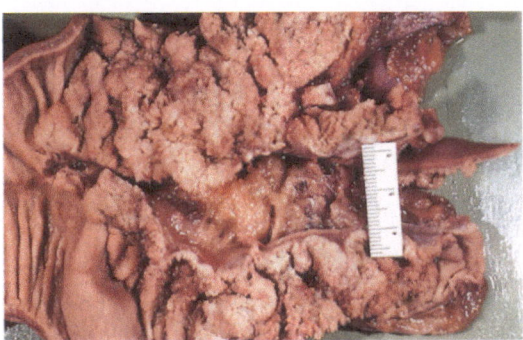

FIGURE 2.2. Large velvety sessile villous adenoma split in half, with extensive lateral spread. Normal mucosa to left. Villous adenomas frequently contain invasive carcinomas and may also represent markers for metachronous carcinomas.

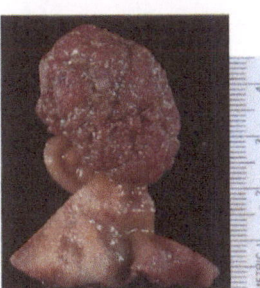

FIGURE 2.3. Tubulovillous adenoma with stalk. Surface is lobulated but also velvety, a feature associated with villosity.

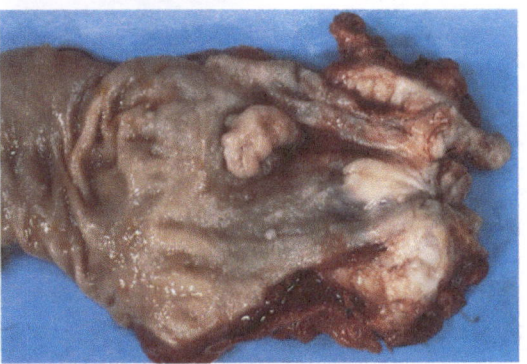

FIGURE 2.17. Malignant polyp. Abdominoperineal resection specimen showing sessile tubulovillous adenoma of distal rectum that contained an 0.8 cm focus of carcinoma extending into submucosa of rectal wall. The two tiny subjacent white sessile polyps were hyperplastic. Two regional lymph nodes were free of carcinoma.

1
Normal Colon

A. Anatomic Considerations

The colon and rectum together compose the large intestine, which extends from the ileocecal valve to the anus. The combined length of the colon and rectum varies considerably. Surgical and anatomic texts indicate an average length of 120–150 cm, whereas at colonoscopy the cecum, in most cases, is approximately 80 cm from the anal verge, after all the loops of bowel have been reduced.

The *cecum* is defined arbitrarily as that portion of the large intestine below the level of entry of the ileum. The ascending colon extends upward from the cecum to the hepatic flexure. The transverse colon loops down toward the midabdomen from the hepatic flexure and then up to the splenic flexure. The descending colon passes downward from the splenic flexure and becomes the sigmoid colon over the medial border of the left psoas major muscle. The sigmoid colon is suspended on a mesentery and forms a loop of variable length. It joins the rectum in front of the third piece of the sacrum. The rectum goes through an abrupt turn to enter the anus 2–3 cm in front of and below the tip of the sacrum.

The degree of peritoneal investment of the large intestine has practical implications for the endoscopist. The sigmoid colon, transverse colon (apart from 7.5–10.0 cm immediately distal to the hepatic flexure), ascending colon, and cecum are completely covered and therefore much more mobile than the remaining portions, which are all at least partly retroperitoneal. The mobile portions are usually the most difficult stretches of the bowel to negotiate with the colonoscope and may therefore account for incomplete examinations.

B. Histology

The mucosa is 0.4–0.6 mm thick and consists of regularly arranged vertical crypts and intervening connective tissue of lamina propria. The crypts, which are lined by one layer of columnar epithelium that is bounded externally by a basement membrane, show a predominance of goblet cells; in the superficial epithelium, absorptive cells are relatively more common. In the base of the crypts are mature endocrine cells as well as undifferentiated and partly differentiated cells that contain free ribosomes, a few rudimentary microvilli, and apical vesicles staining for glycoprotein (mucin) (Lorenzsonn & Trier, 1968). Some of the apical mucin droplets may represent precursors of goblet cell mucins. Above the crypt base are recognizable absorptive and goblet cells that develop from the partly differentiated cells in the crypt bases and whose ultrastructural characteristics have been well documented by Pittman and Pittman (1966) and others. Of interest are the glycocalyceal bodies, round intraluminal structures of unknown significance associated with the surface microvilli, and the glycoprotein-rich apical vesicles (vacuoles), which differ from classical mucous granules: Both features are found in absorptive and goblet cells. It is believed by one group that absorptive cells actually represent mucous cells depleted of secretion and not a separate cell type (Shamsuddin et al., 1982), but this is a minority view. It seems probable that all the specialized epithelial cells of the human colon (i.e., columnar, mucous, and endocrine) are derived from pluripotential stem cells in the deep crypts, as has been shown by Leblond and his colleagues in the mouse (see Kirkland, 1988). The lamina propria contains a scattering of chronic inflammatory cells and lymph follicles that may straddle the muscularis mucosae, extending into submucosa. The epithelium lining crypts in or adjacent to these follicles is basophilic and has a reduced number of goblet cells. It has been postulated that these lymphoglandular complexes and their associated M cells (epithelial cells that can internalize certain luminal proteins) are sites of antigen recognition and processing (O'Leary & Sweeney, 1986); lymphoid follicles are also said to be sites of adenoma formation (Oohara et al., 1981).

Although blood vessels abound throughout the lamina propria, mucosal lymphatics are associated with the muscularis mucosae and never extend above the crypt bases (Fenoglio et al., 1973), a finding that may account for the absence of metastases from APs containing carcinoma in situ, that is, carcinoma confined to the crypts by their basement membranes.

Colonic epithelial cells normally migrate up the crypts to the surface where they are exfoliated, a process taking three to eight days (Lipkin et al., 1963; MacDonald et al., 1964; Eastwood, 1977). The proliferative zone, where cells retain the ability to synthesize deoxyribonucleic acid (DNA), is confined to the lower two thirds of the crypts. An upward shift of the proliferative zone is noted with increasing age, and it has been postulated that this may account for the higher incidence of colorectal neoplasia in the elderly (Roncucci et al., 1988), since such a shift is found in both APs and colorectal cancers. Calcium appears to inhibit proliferation in normal, but not in neoplastic, colon epithelium (Buset et al., 1986). The crypts are surrounded by a sheath of fibroblasts that migrates upward synchronously with the crypt epithelium. During this migration the fibroblasts simultaneously lose their ability to synthesize DNA and acquire the capacity to secrete collagen (Kaye et al., 1971). Recent work with monoclonal antibodies and the electron microscope suggests that at least some of the sheath cells have contractile properties and may be myofibroblasts (Richman et al., 1987).

Although deep crypt cell replication and migration have been extensively investigated, little attention has been paid to crypt regeneration as a whole. It is believed that this is expressed in the form of branching crypts. Although branching is rarely seen normally, it is commonly found in chronic inflammatory bowel disease, presumably as the result of accelerated cell loss (Cheng et al., 1986). Interestingly enough, it is also seen in familial polyposis coli (FPC) despite the absence of crypt damage in that condition. Its presence may reflect defective regulation of crypt production in FPC.

Colonic goblet cells contain both sialomucin and sulfomucin, the proportions varying according to the anatomic segment and the level within the crypt. They also react with the periodic acid–Schiff (PAS) stain. This PAS reactivity varies according to anatomic site within the colon but is generally much weaker than that exhibited by other epithelial mucins (Lev & Spicer, 1965; Lev & Orlic, 1974). These staining reactions undergo changes during neoplastic transformation, which will be discussed in Chap.2.F.3. Heterogeneity of secretory glycoproteins and of goblet cell populations has also been demonstrated using labeled lectins and other immunohistochemical techniques to depict blood group substances and various mucin species in the colonic mucosa (see Podolsky et al., 1986).

References

Buset M, Lipkin M, Winawer S, Swaroop S, Friedman E (1986) Calcium, cellular proliferation and cancer. Cancer Res 46:5426–5430.

Cheng H, Bjerknes M, Amar J, Gardiner G (1986) Crypt production in normal and diseased human colonic epithelium. The Anatomical Record 216:44–48.

Eastwood GL (1977) Progress in gastroenterology: Gastrointestinal epithelial renewal. Gastroenterology 72:962–975.

Fenoglio CM, Kaye GI, Lane N (1973) Distribution of human colonic lymphatics in normal, hyperplastic and adenomatous tissue. Its relationship to metastasis from small carcinomas in pedunculated adenomas, with two case reports. Gastroenterology 64:51–66.

Kaye GI, Pascal RR, Lane N (1971) The colonic pericryptal fibroblast sheath: Replication, migration, and cytodifferentiation of a mesenchymal cell system in adult tissue. III. Replication and differentiation in human hyperplastic and adenomatous polyps. Gastroenterology 60:515–536.

Kirkland SC (1988) Clonal origin of columnar, mucous, and endocrine cell lineages in human colorectal epithelium. Cancer 61:1359–1363.

Lev R, Orlic D (1974) Histochemical and radioautographic studies of normal human fetal colon. Histochemistry 39:301–311.

Lev R, Spicer SS (1965) A histochemical comparison of human epithelial mucins in normal and in hypersecretory states including pancreatic cystic fibrosis. Am J Pathol 46:23–47.

Lipkin M, Bell B, Sherlock P (1963) Cell proliferation kinetics in the gastrointestinal tract of man. I. Cell renewal in colon and rectum. J Clin Invest 42:767–776.

Lorenzsonn V, Trier JS (1968) The fine structure of human rectal mucosa: The epithelial lining of the base of the crypt. Gastroenterology 55:88–101.

MacDonald WC, Trier JS, Everett NB (1964) Cell proliferation and migration in the stomach, duodenum, and rectum of man: Radioautographic studies. Gastroenterology 46:405–417.

O'Leary AD, Sweeney EC (1986) Lymphoglandular complexes of the colon: Structure and distribution. Histopathology 10:267–283.

Oohara T, Ogino A, Tohma H (1981) Microscopic adenoma in non-polyposis coli: Incidence and relation to basal cells and lymphoid follicles. Dis Colon Rectum 24:120–126.

Pittman FE, Pittman JC (1966) An electron microscopic study of the epithelium of normal human sigmoid colonic mucosa. Gut 7:644–677.

Podolsky DK, Fournier DA, Lynch KE (1986) Human colonic goblet cells. Demonstration of distinct subpopulations defined by mucin-specific monoclonal antibodies. J Clin Invest 77:1263–1271.

Richman PI, Tilly P, Jass JR, Bodmer WF (1987) Colonic pericrypt sheath cells: Characterization of cell type with new monoclonal antibody. J Clin Pathol 40:593–600.

Roncucci L, Ponz de Leon M, Scalmati A, Malagoli G, Pratissoli S, Perini M, Chahin NJ (1988) The influence of age on colonic epithelial cell proliferation. Cancer 62:2373–2377.

Shamsuddin AM, Phelps PC, Trump BF (1982) Human large intestinal epithelium: Light microscopy, histochemistry and ultrastructure. Hum Pathol 13:790–803.

2
Pathology

Polyp [Gr. *polypous* a morbid excrescence]
Dorland's Illustrated Medical Dictionary

A. General Prevalence and Anatomic Distribution

Because they are usually asymptomatic, the incidence rate of polyps is difficult to evaluate directly. The prevalence of polyps is a useful, although indirect, measure of their incidence.

As indicated in the Introduction, the prevalence of polyps (this will refer in the following account to APs) depends on how, when, and on whom the pertinent studies were conducted. Thus, Rider et al. (1954) found a prevalence of 5.1 percent in his sigmoidoscopic survey of symptomatic US subjects whose average age was 52 years, whereas autopsy studies of (older) US subjects have revealed rates up to 69 percent. The typical prevalence in autopsy series is 10–50 percent (Clark et al., 1985): It is higher when a hand lens is used to inspect the mucosa. Marked geographic variation is noted. The prevalence is highest in the United States, somewhat lower in western and northern Europe, and lowest in eastern Europe and developing countries (Eide & Stalsberg, 1978; A.R. Williams et al., 1982). Representative figures are shown in Table 2.1.

In determining the location of polyps, a clear definition and measurement of the various colonic segments are necessary. Thus, in their 1979 study Rickert et al. defined the *cecum* as the first 5 percent of the colon, the *ascending colon* as the subsequent 6–20 percent, the *transverse colon* as 21–50 percent, the *descending colon* as 51–65 percent, the *sigmoid* as 66–90 percent, and the *rectum* as the terminal 91–100 percent. A similar system was used by Spratt et al. (1958) in their surgical series and by A.R. Williams et al. (1982) and by Eide and Schweder (1984) in their autopsy studies. In this fashion the number of polyps found in each segment can be compared with the number expected per unit length of colon. The *left* (or *distal) colon* is defined by most investigators as colon distal to the splenic flexure. It has been found by several observers using this method that polyps at autopsy are fairly evenly distributed throughout the colon or that there may even be a right-sided predominance (Stemmermann & Yatani, 1973; Rickert et al., 1979). The latter study, however, involves bias, owing to exclusion of subjects with previously diagnosed carcinomas, most of which were located distally. This even distribution contrasts with most colonoscopic surveys, which have found a

TABLE 2.1. Prevalence of adenomas and incidence of colorectal carcinoma in different countries.

	Prevalence of adenomas (%) > 50 yr	Incidence of carcinoma per 100,000 (persons 36–64 yr)
South African Bantu	0	Low
Southern Iran	2	Low
Columbia (Cali)	12.7 (> 45 yr)	10.9
Japan (Miyagi)	16.4	17.3
Norway (Oslo)	36.3	26.0
England (Liverpool)	37.2	42.1
United States (New York State)	51.2	49.1
Hawaii (Japanese)	66.7	48.9

SOURCE: Slightly modified from Morson et al., 1983, and Berg, 1988.

left-sided predominance of polyps (e.g., Gillespie et al., 1979; Shinya & Wolff, 1979; Frühmorgen & Matek, 1983; Morson et al., 1983; Isbister, 1986; Wegener et al., 1986). Some of these studies, on the other hand, were performed on symptomatic patients in whom one would expect to find larger polyps, a factor introducing its own bias (see B.2 below). This apparent discrepancy between autopsy studies, generally performed on more elderly subjects, and colonoscopic and surgical surveys, which generally deal with younger patients, can at least partly be explained by the fact that polyps appear first distally and are joined in subsequent years by more proximal polyps (Eide & Stalsberg, 1978; Granqvist, 1981; Eide & Schweder, 1984). Consistent with this hypothesis is the observation at autopsy that left-sided polyps are larger than those on the right, suggesting a longer duration of the former lesions. Also of interest in this regard is the fact that a relatively higher percentage of polyps has been found in the right colon in the past two decades, especially in the United States (Greene, 1983; Webb et al., 1985). This right-sided increase parallels that found for colorectal carcinomas and may partially result from aging of the population. Both shifts have potential public health significance concerning screening for and treatment of colorectal neoplasms.

B. Gross Appearance

Adenomatous polyps, which compose tubular, villous, and tubulovillous adenomas (Morson & Sobin, 1976), are protrusions of mucosal epithelial tissue into the colonic lumen. Although macroscopic differences have been described between adenomatous and hyperplastic polyps (the pallor of the latter is frequently contrasted with the redness of the former), most endoscopists cannot consistently differentiate grossly between the two lesions, especially the smaller ones (Granqvist et al., 1979; Isbister, 1986; discussion in Winawer et al., 1988). In an attempt to quantify this, Neale et al. (1987) compared the diagnostic accuracy of three experienced endoscopists using the flexible sigmoidoscope on

718 asymptomatic patients. They found that 80 percent of the polyps were diagnosed accurately and that the errors were highest for adenomatous polyps under 0.5 cm and for hyperplastic polyps over 0.5 cm. A similar result (85 percent accuracy) was reported by Chapuis et al. (1982) in a study of symptomatic individuals.

The majority of adenomatous polyps are tubular adenomas that consist of smooth-surfaced, lobulated masses having a raspberrylike appearance (Figure 2.1, appears in color preceding page 1).

They may be pedunculated (stalked) or sessile (broad based). The percentage of polyps that are pedunculated is higher in the larger polyps and in left-sided, especially sigmoid, lesions; typical figures range from 41 percent in the autopsy study of Eide and Stalsberg (1978) to 68 percent in the surgical study of Grinnell and Lane (1958). Some workers include a short-stalked or semipedunculated group in their classification schemes.

A velvety or papillary appearance is suggestive of villous adenoma (Figure 2.2, appears in color preceding page 1). Villous adenomas are more frequently sessile and have a less well circumscribed field of origin than tubular adenomas and may have satellite growths; because of these features, they may recur after excision (Day & Morson, 1978). Tubulovillous adenomas show an admixture of lobular and papillary areas (see Figure 2.3 and Figure 5-3 in Day & Morson, 1978).

An attempt to refine the gross characteristics of adenomatous polyps was provided by Thompson and Enterline (1981). They stained polyps removed colonoscopically with 1 percent trypan blue and examined them with a dissecting microscope at ×10–30. They developed macroscopic criteria for differentiating tubular from villous adenomas based on the configuration and number of lobules. They were also able to predict the grade of dysplasia and the presence of invasive carcinoma, based on the degree of irregularity of the gyri and on the presence of mucosal depressions or ulcerations. They felt that these observations could be helpful in the selection of (1) APs for frozen section (a rarely performed procedure, however) and (2) mucosal sites in patients with ulcerative colitis most likely to show severe dysplasia, which would be an improvement over the blind biopsies currently performed. A similar study of minute lesions performed by Nishizawa et al. (1980) using ×30 magnification revealed disturbances in the pattern of the colonic pits, which they felt were characteristic of early adenomatous change, early carcinoma in situ, or carcinomatous change within adenomatous polyps.

The ability to identify grossly foci of significant dysplasia in a polyp could be important in view of the problem of how to manage the diminutive (< 0.5 cm) polyp (Chaps. 2.E and 5.C.4).

1. Multiplicity

The percentage of polyp-bearing subjects showing multiple polyps ranges from 12 percent (Rider et al., 1954) to slightly over 50 percent. The 12 percent figure, however, represents the results of a sigmoidoscopic study supplemented by some radiologically detected lesions and is surely an underestimate of the true percentage: The same authors found only a 5.4 percent overall prevalence of polyps, a

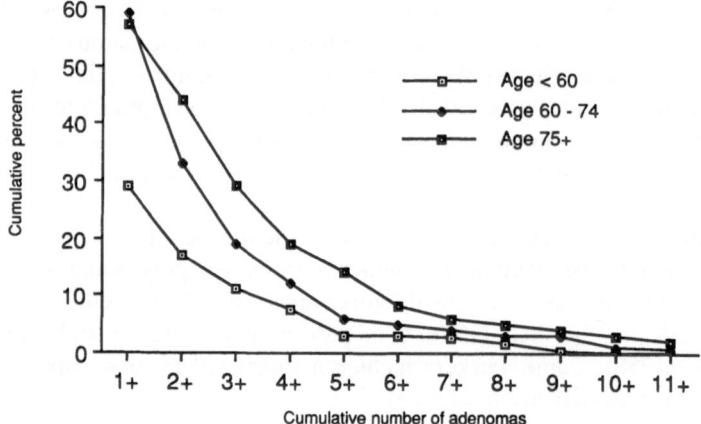

FIGURE 2.4. Cumulative percentage of adenomas by age group. (From Rickert et al., 1979.)

figure well below that found in similar populations by other investigators. One of the highest figures, namely the 53 percent of multiplicity found by Isbister (1986), is ascribed by that author to the high rate of colorectal carcinoma in New Zealand. Current data indicate that on the average 40–50 percent of patients with one AP will have additional polyps.

The percentage of patients bearing multiple polyps increases with age (Figure 2.4). For example, Rickert et al. (1979) found a 14 percent incidence of two polyps under age 60 and a 44 percent incidence over age 75. Multiplicity is also higher when there is an accompanying colorectal carcinoma. Grinnell and Lane (1958), for example, found multiple polyps in 34 percent of cases with a coexistent carcinoma and in 17 percent of cases without carcinoma.

Multiple polyps are significant since they represent a clear marker for the presence of synchronous and metachronous colorectal carcinomas and are thus one of the most important risk factors to be considered when determining post-polypectomy surveillance schedules.

2. Size

Polyp size is, again, largely dependent on the study population and on polyp histology (tubular versus villous). Autopsy studies, especially those excluding earlier polypectomies or colectomies for neoplasia, have a lower mean size than that found in surgical series that deal with symptomatic individuals or with patients having separate colorectal carcinomas. Thus, the mean polyp sizes in the autopsy studies of Rickert et al. (1979) and of A.R. Williams et al. (1982) were 5.8 mm and 5.6 mm, respectively, whereas it was 1.6 cm in the surgical series of Grinnell and Lane (1958). Villous adenomas are larger than tubular adenomas. In a representative surgical series, a mean size of 1.2 cm was recorded for benign tubular adenomas and 3.7 cm for benign villous adenomas (Grinnell & Lane, 1958).

The relationship of size to patient age merits some discussion. The overall mean size of polyps generally shows no effect of age. The percentage of polyps > 1 cm does increase with age, however (Eide & Stalsberg, 1978; Rickert et al., 1979; A.R. Williams et al., 1982): The latter noted that 4.6 percent of subjects under 54 years of age fell into that category, whereas 15.6 percent over 75 years did. Granqvist (1981) found this increase in size with advancing age to be most marked proximal to the sigmoid colon and felt that this supported the idea that new polyps appear mainly in the right colon and that they then grow; but this belief has been challenged by others.

It is important to record the size of polyps. Large size is a risk factor not only for carcinoma in that polyp but also for metachronous carcinomas (C.B. Williams & Macrae, 1986; Winawer et al., 1986).

C. Microscopic Features

1. Tubular versus Villous Adenomas

Well-developed *tubular adenomas* consist of branching tubules lined by closely packed cells containing elongated nuclei parallel to the long axis of the cells and basophilic cytoplasm with reduced amounts of mucus. *Villous adenomas* are composed of elongated papillae lined by a similar epithelium. In classical villous adenomas the villi extend in an uninterrupted manner from the muscularis mucosae to the luminal surface. *Tubulovillous adenomas* contain admixtures of tubules and villi. Not infrequently, villi are found on the surface of the polyp and tubules in the deeper areas.

There are currently various classification schemes. Konishi and Morson (1982) require 80 percent of a polyp to be villous before designating it as a villous adenoma, whereas Spjut and Appel (1979) only require 50 percent. The former classify polyps with 20–80 percent villi as tubulovillous, whereas the latter prefer to designate polyps with < 50 percent villi as tubular, followed by an estimate of the percentage of villi.

Classification of polyps as tubular, villous, or tubulovillous is fraught with difficulties not only because of the lack of standardized quantitative criteria, as discussed above, but also because of the frequently uneven histological sampling of the polyps. In addition, it may be difficult to differentiate "true" villi from "apparent" villi between deep crypts on the surface of a tubular adenoma or between invaginating and irregular clefts within lobules of adenomatous tubules (see Figures 2.5–2.7 and Chap. 2.G on the histogenesis of polyps). In fact, deeper sections through lobules such as those depicted in Figure 2.5, thereby obtaining a three-dimensional view of the polyp, may reveal definite villi that had not been apparent in the initial sections. The result is considerable variation in the

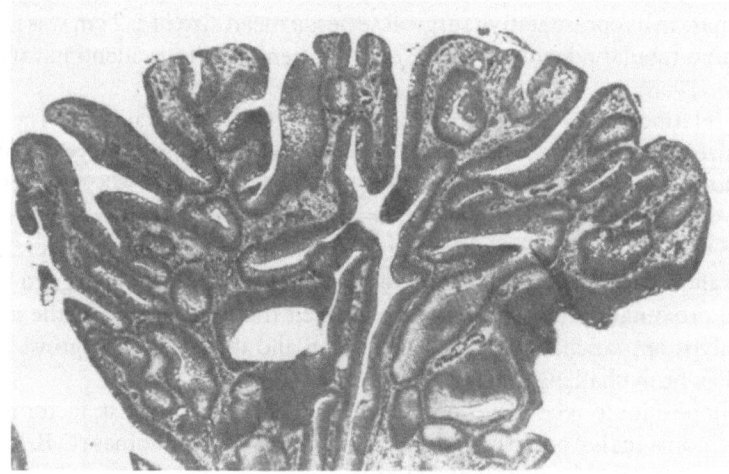

FIGURE 2.5. Small adenomatous polyp showing lobulations partially subdivided by clefts. Apparent villi are seen on the surface on the left. H&E; ×44, reproduced at 100%.

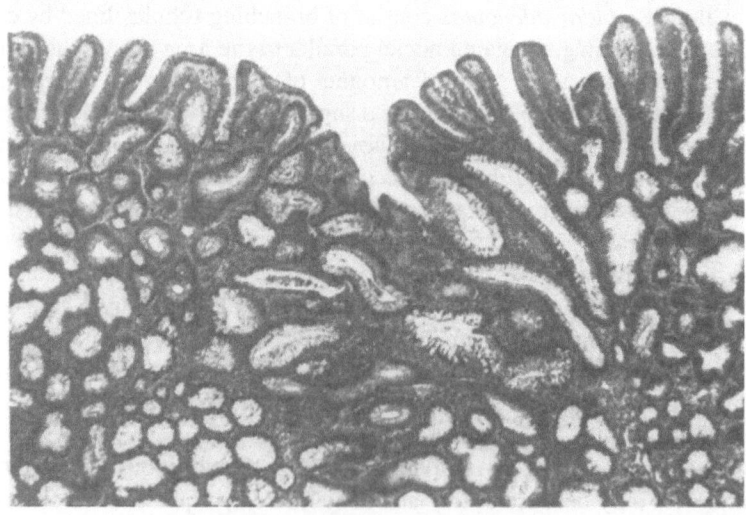

FIGURE 2.6. Surface of adenomatous polyp showing early villous formation on right and presumed earlier stage of same process on upper left. The bulk of the polyp consists of adenomatous tubules. H&E; ×44, reproduced at 100%.

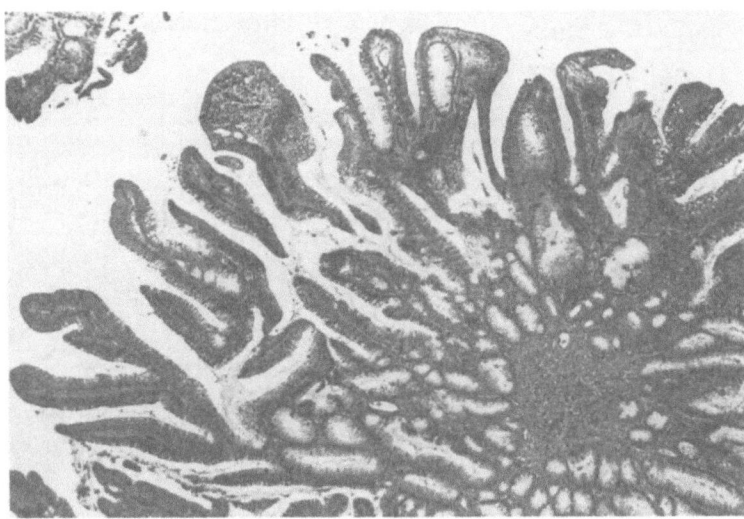

FIGURE 2.7. Tubulovillous adenoma. Well-developed branching villi are seen on left and presumed earlier stage of formation of villi toward right. H&E; ×44, reproduced at 100%.

proportions of the three lesions in different series. In one fairly representative series of 1,187 adenomas, 81 percent were classified as tubular, 16 percent as tubulovillous, and 3 percent as villous (Konishi & Morson, 1982), whereas in another series of 4,701 adenomas 72 percent, 22 percent, and 6 percent, respectively, were so classified (Hermanek et al., 1983). In virtually all series, however, tubular adenomas constitute the vast majority of polyps. The percentage of villous adenomas (VAs) and tubulovillous adenomas (TVAs) is higher in the rectosigmoid colon than in the right colon. An accurate estimate of the villous component is important since villosity constitutes one of the risk factors for carcinoma in that polyp and possibly for metachronous carcinoma.

2. Degree of Dysplasia

Progressive dysplasia is characterized by increasingly severe architectural and cytological abnormalities. Of special interest are the early architectural changes in the mildly dysplastic (well-differentiated) polyps. Initially, the glands undergo slight enlargement and elongation. The proliferation of the adenomatous cells eventually exceeds the capacity of the adenomatous tubules to accommodate them. The nests of proliferating cells may then indent the basement membrane with minute secondary lumen formation and may then bulge into the stroma, or they may produce papillary excrescences into the lumen. Continued proliferation then results in extensive buckling of the basement membranes (Figures 2.8–2.10).

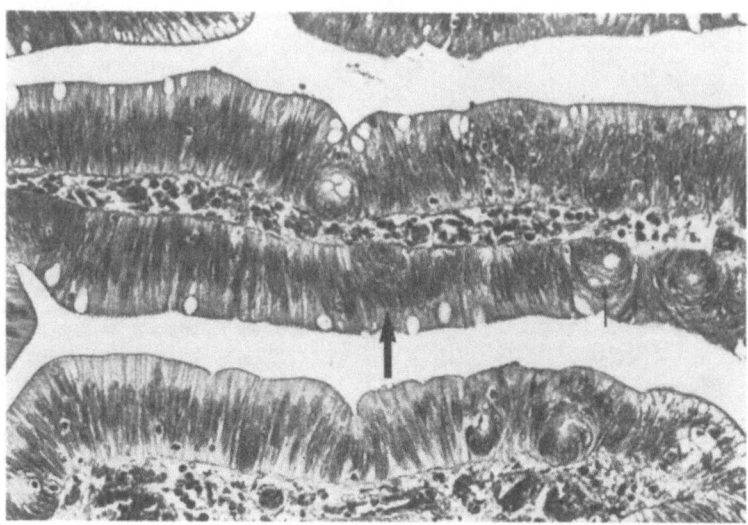

FIGURE 2.8. Strips of adenomatous epithelium from adenomatous polyp showing prolifer-ating cells within epithelial layer (large arrow), some of which have slightly indented the basement membrane or formed early secondary lumina (small arrow). H&E, ×282, reproduced at 100%.

FIGURE 2.9. Adenomatous glands showing proliferating cell clusters with secondary lumina. The budding is slightly more advanced than in Figure 2.8. H&E; ×88, reproduced at 107%.

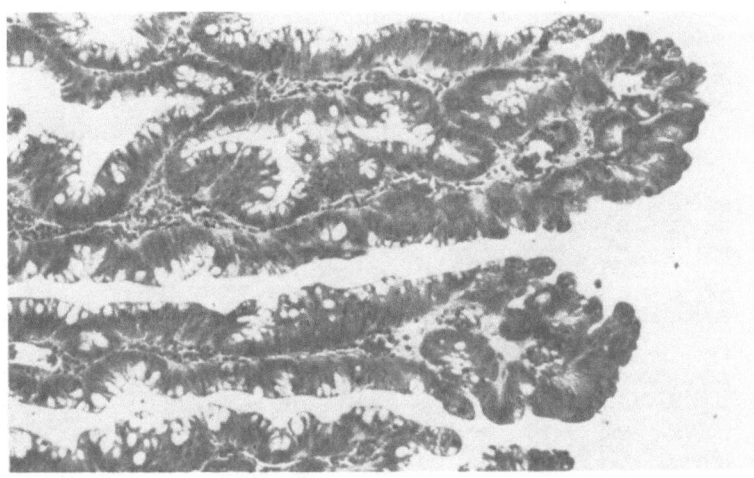

FIGURE 2.10. Adenomatous tubules showing extensive buckling of epithelium with cell clusters extending into lumen and into stroma. H&E; ×88, reproduced at 100%.

This is followed by further expansion of adjacent stromal nests, which may impart a "back-to-back" appearance to these glands, and by proliferation and bridging of intraluminal cells, which results in secondary lumen formation and a cribriform pattern (Figures 2.11 and 2.12). Cytological changes accompanying these architectural characteristics of increasing dysplasia include loss of nuclear polarity, nuclear enlargement, and increase in the nuclear/cytoplasmic ratio, in nuclear pleomorphism, and in the numbers of mitoses. Other features include reduction in cytoplasmic mucin, a common event, and "goblet cell reversal," that is, mucin droplets in a basal position, which occurs less frequently.

There are several ways of classifying dysplasia in polyps. Three grades are used by Morson and colleagues (Konishi & Morson, 1982). In that system, all APs by definition show dysplasia, which is categorized as mild, moderate, or severe. In their series of 1,187 APs, 81.9 percent showed mild, 11.6 percent moderate, and 6.5 percent severe dysplasia (Morson et al., 1983). Several Japanese groups use a five- or six-grade system. According to Kozuka (1975), the degree of epithelial pseudostratification (position of the nuclei in the cell) and the pattern of glandular branching are of critical importance in grading dysplasia, correlating well with and supplementing the above criteria. His grades 0–III, which correspond to mild dysplasia, show progressive gland enlargement, shift of nuclei from perpendicular to parallel to the long axis of the cell, slight cell crowding, and mild pseudostratification. I have found that these early changes may be difficult to differentiate from reactive crypts in mildly inflamed or irritated mucosae (see Figures 2.13, 2.14, and 4.4).

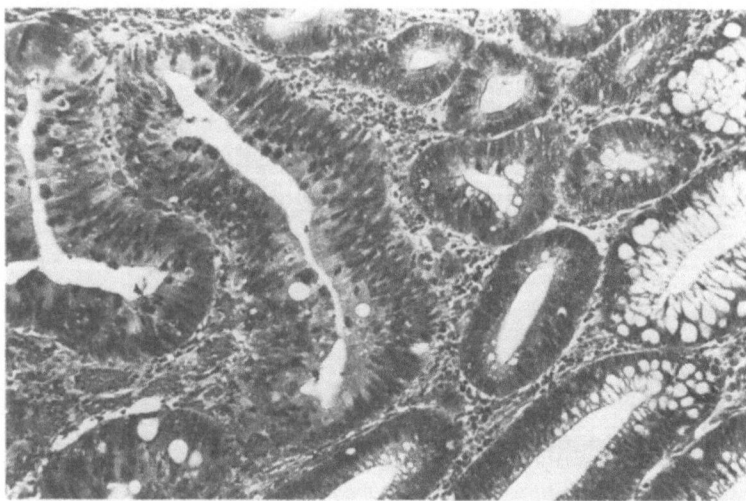

FIGURE 2.11. Adenomatous polyp with varying grades of dysplasia in same field. Well-differentiated (mildly dysplastic) glands on extreme right have basal nuclei and preserved mucus. Glands in right center show partial loss of nuclear polarity and marked reduction in mucus. The more severely dysplastic glands on the left show mild architectural abnormalities (irregular shapes) as well as cytological changes. H&E; ×176, reproduced at 55%.

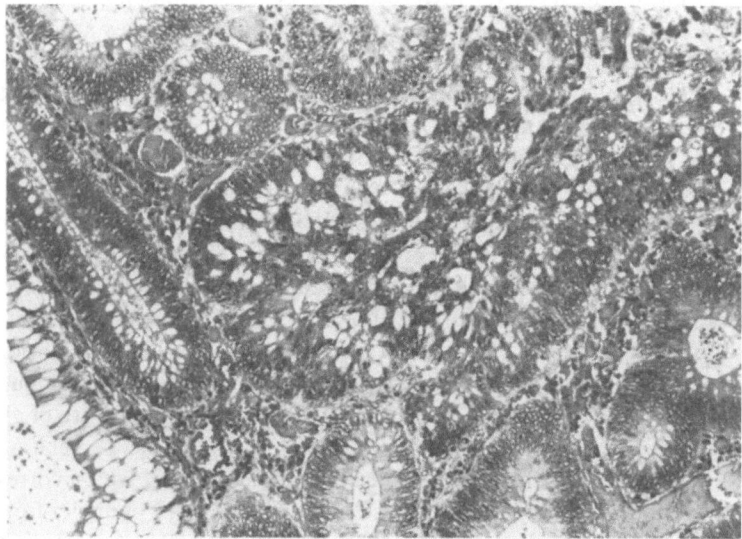

FIGURE 2.12. Varying dysplasia in same field. Crypt in lower left contains mildly reactive (trapped) epithelium. Glands along bottom and top show mild to moderate dysplasia. In the upper center they merge with the central sheet of severely dysplastic epithelium. H&E; ×176, reproduced at 55%.

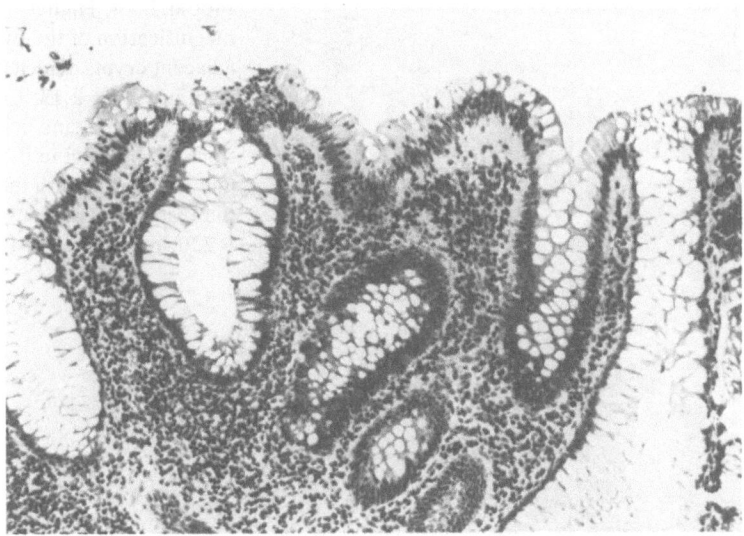

FIGURE 2.13. Mucosal excrescence showing inflammation and reactive crypts. Fairly normal crypt on far right has small basal nuclei that are parallel to the lumen. The crypts to its left contain cells with numerous large nuclei that are perpendicular to the lumen. Such reactive changes mimic, and may be difficult to differentiate from, those found in early, mildly dysplastic adenomas. H&E; ×176, reproduced at 55%.

The more advanced of these early changes have been called *preadenomatous* by Lev et al. (1987). In Kozuka's grade IV, equivalent to moderate dysplasia, there is severe pseudostratification, but the most apical cytoplasm is still devoid of nuclei, whereas in grade V, severe dysplasia, nuclei are found there. I have found that the position of nuclei as outlined above provides supplementary help in grading dysplasia but that the branching pattern, which depends on the degree of primary and secondary branching of glands, is less useful. In a quantitative study, Taki (1980) was able to correlate increasing dysplasia with increasing height of epithelial cells and with increasing nuclear volume of the cells.

Adenomatous epithelium, especially that located on the colonic luminal surface, may react as does nonneoplastic epithelium to erosion, inflammation, and irritation, with resultant exaggeration of the degree of dysplasia. Such changes include enlarged and atypical nuclei, nuclear stratification, and surface papillations. (See Figure 2.15.)

Islands of benign squamous epithelium, generally designated as *squamous metaplasia*, may also be found (Figure 2.16). In our series of 229 APs, we found 1 such case (0.4 percent). Similar islands have been identified by others in 0.4 to 0.9 percent of APs (G.T. Williams et al., 1979; Sarlin & Mori, 1984). They may represent precursors of squamous cell carcinomas of the colon, very rare lesions that are found predominantly in the proximal colon.

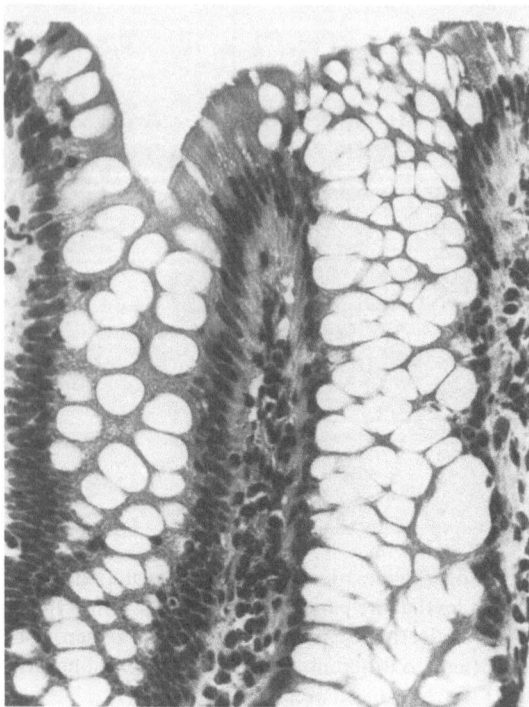

FIGURE 2.14. Higher magnification of the two adjacent crypts depicted in far right in Figure 2.13. Contrast the size, shape, and orientation of the nuclei in the normal right crypt with those in the reactive left crypt. H&E; ×220, reproduced at 100%.

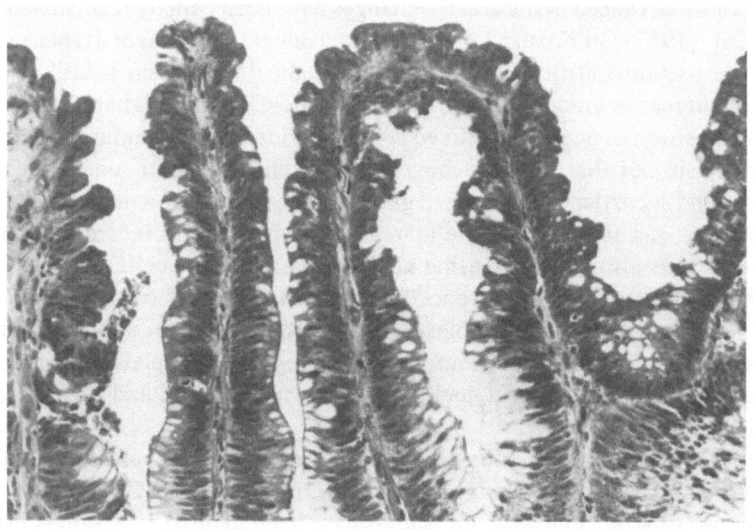

FIGURE 2.15. Surface of adenomatous polyp. Luminal areas show papillary tufts, disordered cell maturation with loss of nuclear polarity, and reduced mucus. These represent reactive changes in the mildly dysplastic epithelium shown in more classic form in subjacent areas, rather than a higher grade of dysplasia. H&E; ×282, reproduced at 55%.

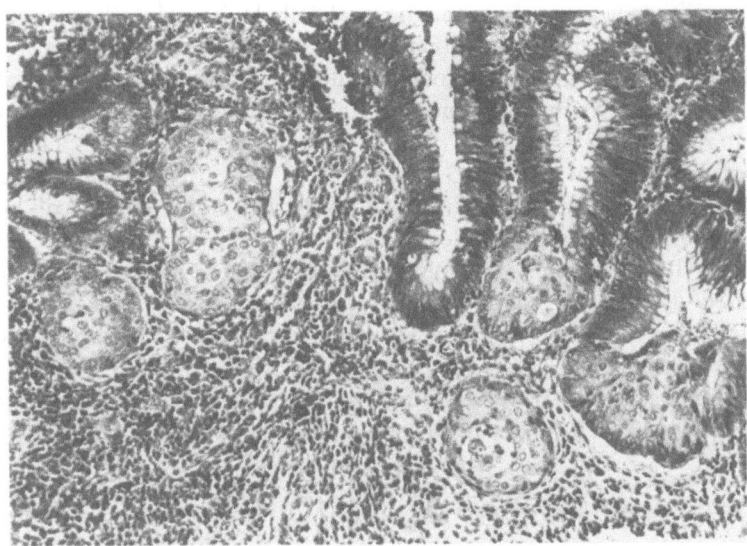

FIGURE 2.16. Squamous metaplasia in adenomatous polyp. The metaplastic cell clusters contain abundant pale eosinophilic cytoplasm. H&E; ×176, reproduced at 55%.

a. Problems in Grading Dysplasia

In most polyps, varying degrees of dysplasia can be found. Both Kozuka (1975) and Konishi and Morson (1982) believe that the assigned grade should reflect the most dysplastic changes even though these may be quantitatively insignificant. I have found that it may not be easy to apply this principle. Does one crypt exhibiting the higher grade justify placing the entire polyp in that category, or are several clusters of such crypts or a minimum percentage required? Similar difficulties in establishing the minimal size of involvement were experienced by a group of gastrointestinal pathologists attempting to grade dysplasia in Barrett's esophagus (B.J. Reid et al., 1988). Perhaps of even greater importance is the occasional inability to grade the changes at all. Frequently, the grades merge imperceptibly with one another, and it may be difficult to identify with certainty the highest grade present. Moreover, I have found that some of the characteristics of moderate dysplasia (e.g., mild variation in nuclear size and mild nuclear stratification) may be confused with reactive changes in mildly dysplastic surface epithelium as described and illustrated above. (Nuclear stratification in surface glands in Barrett's epithelium was interpreted variously as indefinite for dysplasia, low-grade dysplasia, and high-grade dysplasia in the Reid study cited above.)

Lack of uniform criteria may account for the wide variations in polyp grade even among experienced gastrointestinal pathologists. For example, moderate dysplasia was found in 22 percent of APs by Muto et al. (1975) but only in 7 percent of polyps by the pathologists in the National Polyp Study (NPS) (Winawer et al., 1988). The NPS pathologists also noted frequent disagreement of one grade amplitude between them and the local pathologists submitting cases to that study (see M. O'Brien in discussion in Winawer et al., 1988). Brown et al. (1985) found substantial inter- and intraobserver disagreement in the histological grading of dysplasia. Their morphometric analysis showed that (1) nuclear/cytoplasmic ratio and (2) variability (standard deviation) of nuclear area and height (but not absolute nuclear area and height) were of value in discriminating between the three grades of dysplasia, but it is unlikely that morphometry will be commonly used in diagnostic pathology.

Despite these problems in histological interpretation, it is necessary to grade accurately dysplasia in polyps since severe dysplasia is a recognized risk factor for carcinoma in that polyp and is also associated with separate (synchronous) colorectal carcinoma (Kalus, 1972; Konishi & Morson, 1982). Some workers also consider severe dysplasia as a marker for metachronous carcinoma, whereas others do not (C.B. Williams & Macrae, 1986; Winawer et al., 1986).

3. Carcinoma in a Polyp

It is best to refer to carcinoma in a polyp only when there is clear penetration of carcinoma through the muscularis mucosae into the submucosa, that is, invasive carcinoma (see Figure 2.17 and Frontispiece, both appear in color preceding page 1). This is also called *malignant polyp*, whose behavior and management will be fully discussed in Chap. 5.B. When malignant changes are confined to the glands and do not penetrate the epithelial basement membrane, the lesion is called *severe dysplasia*. Some pathologists also call this *carcinoma in situ*, but this term is avoided by other pathologists because of fear that unnecessary surgery may result from misinterpretation of such a report. Since such tumors never metastasize, polypectomy is adequate treatment. Transitions between severe dysplasia and invasive carcinoma have been described such as (1) intramucosal carcinoma, where the malignant glands have penetrated the epithelial basement membrane into the mucosal lamina propria (see Enterline, 1976; Fenoglio-Preiser, 1985) and frequently elicit a stromal response; and (2) a slightly more invasive tumor whose glands have extended into but not beyond the thickened and splayed out muscularis mucosae. Little is known about the metastatic potential of these latter lesions, but it is probably very low.

4. Morphometric Studies

Several groups have compared normal to neoplastic mucosa using these techniques. In one such micromorphometric study, based on the position of nuclei in the cells, the ratio of apical cytoplasm ("displacement area") to the area of the entire gland, and nuclear area, the authors were able to discriminate between normal, adenomatous, and carcinomatous glands (Graham et al., 1988). In another study (Nakayama et al., 1988), the rise in nuclear to cytoplasmic ratio from normal mucosa to moderate dysplasia was mainly due to nuclear crowding, whereas the increase in that ratio from severe dysplasia to carcinoma resulted primarily from nuclear enlargement. Similar morphometric and stereological techniques have been used to supplement histological analysis in the grading of dysplasia in ulcerative colitis (Chap. 4.F) and in studies of the histogenesis of tubular and villous adenomas (Chap. 2.G).

D. Differential Diagnosis

1. Hyperplastic (Metaplastic) Polyps

Hyperplastic (metaplastic) polyps are common lesions that can be confused grossly with adenomatous polyps (Table 2.2). They were recognized as non-neoplastic and first termed *hyperplastic* early in this century (Feyrter, 1929). They are pale, generally sessile growths most common in distal colon and rectum (Figure 2.17). Most are < 0.5 cm, only 0.9 percent being > 1.0 cm (G.T. Williams et al., 1980). They frequently increase in number with age. They may be single or multiple, and they may occur alone or with colorectal carcinoma. When substantial numbers are present (a minimum of 50 is required by some workers), one may call the condition *hyperplastic polyposis*, a rare entity (Spjut & Estrada, 1977; G.T. Williams et al., 1980; Bengoechea et al., 1987). Benign extension into submucosa, also called *inverted hyperplastic polyp* (HyP) (Sobin, 1985), may occur. Histologically the upper crypts show papillary infoldings consisting of enlarged bland-looking absorptive and goblet cells. The deep crypts are lined by

TABLE 2.2. Histological classification of polyps from Erlangen Registry of Colorectal Polyps, 1978–1982.

	Number	(%)
Types of polyps (n = 6,378)		
Neoplastic	4,759	(74.6)
Nonneoplastic	1,614	(25.3)
Unclassified	5	(0.1)
Subdivisions of neoplastic polyps (n = 4,759)		
Epithelial	4,715	(99.1)
Adenoma	4,280	(89.9)
Adenoma with adenocarcinoma	421	(8.9)
Adenocarcinoma	14	(0.3)
Others, including lipomas (0.4%), leiomyomas (0.3%), carcinoid tumors (0.1%), and other rare tumors (0.1%)	44	(0.9)
Subdivisions of nonneoplastic polyps (n = 1,614)		
Hyperplastic	1,485	(92.0)
Juvenile	52	(3.2)
Inflammatory	46	(2.9)
Peutz-Jeghers	19	(1.2)
Other and unclassified	12	(0.8)

SOURCE: Modified from Hermanek, 1985.

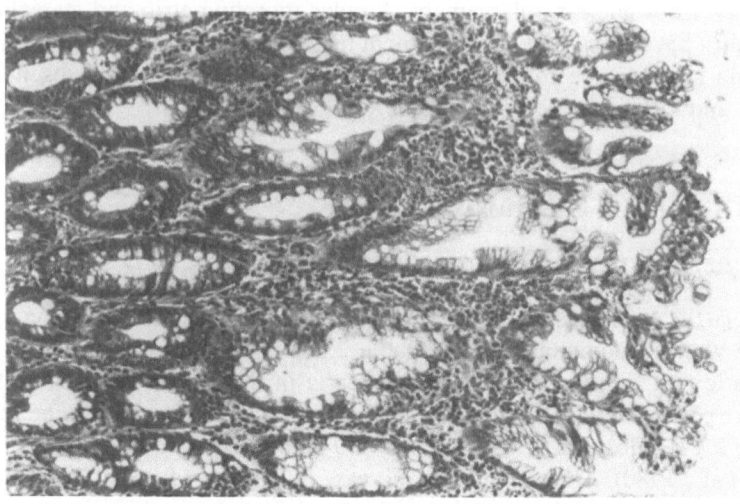

FIGURE 2.18. Hyperplastic polyp. Lumen on right. The upper crypts are dilated and show infoldings of eosinophilic cytoplasm with scattered goblet cells. Deep crypts on left are more basophilic and contain fewer goblet cells than normal. H&E; ×176, reproduced at 53%.

basophilic epithelium that contains reduced numbers of goblet cells and that resembles regenerative epithelium (Figure 2.18).

Ultrastructurally (Kaye et al., 1971, 1973; Hayashi et al., 1974), the cells at each level of the crypt are hypermature, resembling those found at higher levels in normal crypts. Kinetic studies (Kaye et al., 1971; Lane et al., 1971; Hayashi et al., 1974) show that the cells in the lower halves of the crypts normally incorporate tritiated thymidine but migrate more slowly upward than normal cells; superficial epithelium, like the normal, fails to incorporate isotope. Parallel changes are found in the fibroblastic sheath surrounding the epithelium, whose cells show delayed migration and overproduction of collagen beneath surface epithelium. These findings contrast with adenomatous polyps, which do not show normal epithelial or fibroblast maturation and which exhibit persistent DNA synthesis in surface cells. The nature and significance of hyperplastic polyps are unknown. Arthur (1968) stated that they may be hamartomatous, result from response to inflammation/trauma, or represent "degenerative" changes in the mucosa, but favored the latter possibility. They can be regarded as disorders of cell maturation of unknown origin involving deep and upper crypts. Hyperplastic foci identical to those seen in HyP may also occasionally be found in inflammatory and proliferative lesions of the colonic mucosa.

a. Mixed Hyperplastic-Adenomatous Polyps

Some polyps contain hyperplastic and adenomatous components (see Figures 2.19–2.22).

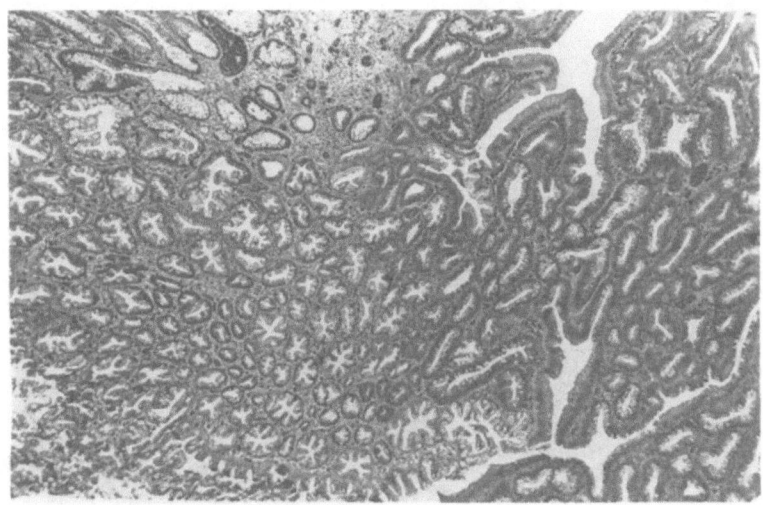

FIGURE 2.19. Mixed hyperplastic-adenomatous polyp. Adenomatous epithelium is seen on the right, and stellate glands typical of hyperplastic polyp are seen at bottom and on the left. H&E; ×22, reproduced at 100%.

The incidence of hyperplastic changes in adenomatous polyps ranges from 0.6 percent (G.T. Williams et al., 1980) to 16 percent (Goldman et al., 1970), and as many as 13 percent of hyperplastic polyps are said to contain adenomatous foci (Estrada & Spjut, 1980). One group originally described gradual transitions between the two epithelia and hypothesized that some APs arose from HyPs (Urbanski et al., 1984) but subsequently reversed themselves and claimed it was more likely that the APs showed secondary hyperplastic changes (Urbanski et al., 1986), a suggestion made earlier by Lane and Lev (1963). Although transitions occur (see Figures 2.20–2.22), they are exceedingly rare. The difficulty in interpreting coexisting changes arises from two sources: (1) erroneous identification of the basophilic glands in deep crypts of HyP as adenomatous and (2) designation of secondary papillary changes in APs as hyperplastic. In APs, papillary ingrowths showing crowded elongated nuclei parallel to the long axis of the cells and basophilic cytoplasm are more likely to be adenomatous than hyperplastic. In the experience of the author, when true transitions do occur, the hyperplastic component is usually above the adenomatous portion of the crypt (e.g., Figures 2.21 and 2.22). This confirms the point that adenomatous epithelium may occasionally show secondary hyperplastic changes rather than the progressively more severe dysplasia which usually characterizes its upward migration.

2. Juvenile Polyps

Juvenile polyps, which predominate in the young, may be single or multiple (Roth & Helwig, 1963). The polyps are smooth surfaced and contain cystically dilated glands in an expanded, often inflamed stroma. The surface epithelium may be

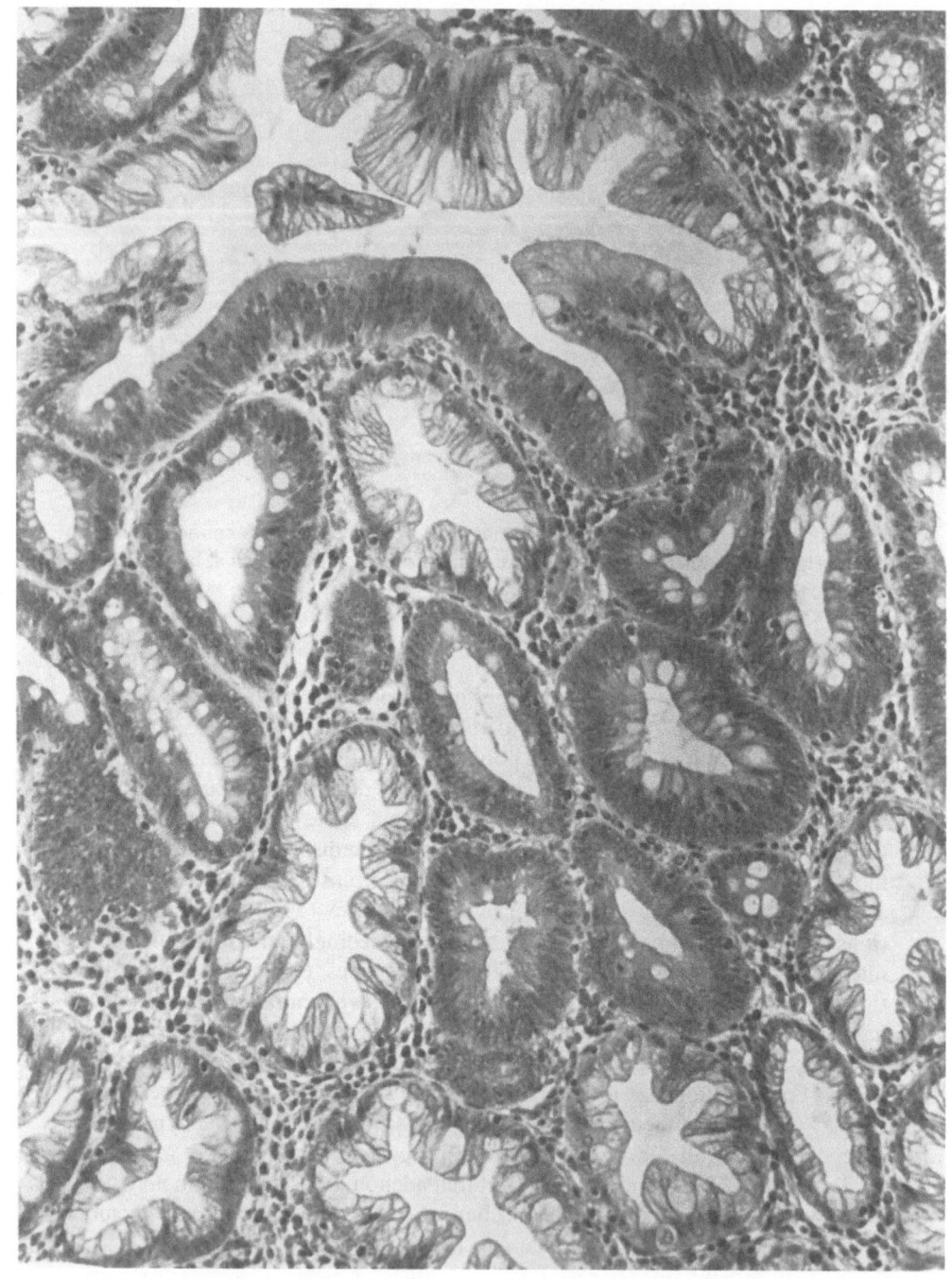

FIGURE 2.20. Higher magnification of mixed polyp shown in Figure 2.19. Adenomatous glands with crowded elongated nuclei contrast with the hyperplastic glands showing cytoplasmic infoldings and smaller, basal nuclei. Large gland on top shows transitions between the two epithelia. H&E; ×88, reproduced at 100%.

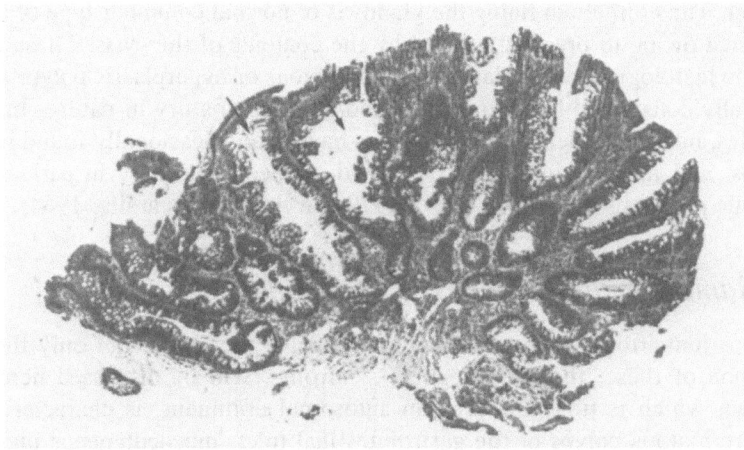

FIGURE 2.21. Mildly dysplastic tubular adenoma (typical glands best seen in lower left and far right) showing changes in several upper crypts (arrows) resembling those found in hyperplastic polyps. H&E; ×22, reproduced at 100%.

FIGURE 2.22. Higher magnification of crypts with arrows from Figure 2.21. The upper crypts show infoldings of eosinophilic cytoplasm (long arrows) that are similar to those found in classical hyperplastic polyps. Papillary infoldings in the deeper crypts of more basophilic cells (short arrows) showing some loss of nuclear polarity probably represent proliferation of truly adenomatous epithelium and contrast with the smooth-walled deep crypts in hyperplastic polyps. (Contrast with Fig. 2.18.) H&E; ×88, reproduced at 55%.

eroded. The epithelium lining the glands is of normal columnar type or may be flattened owing to pressure exerted by the contents of the cysts. These lesions bear no histological resemblance to adenomatous or hyperplastic polyps and are generally considered hamartomatous and/or inflammatory in nature. In recent decades, however, adenomatous changes have been occasionally found in these polyps, and rare carcinomas have been described, especially in patients with juvenile polyposis (Goodman et al., 1979; Jarvinen & Franssila, 1984).

3. Hamartomatous Polyps

Various hamartomatous conditions may involve the colon, but only the most common of these, the Peutz-Jeghers syndrome, will be discussed here. This disease, which is transmitted as an autosomal dominant, is characterized by hamartomatous polyps of the gastrointestinal tract, mucocutaneous pigmentation, and a variety of extraintestinal disorders (Dormandy, 1957). The polyps predominate in the small intestine but may also be found in the colon. They consist of irregular aggregates of nonneoplastic glands separated by smooth muscle bundles. Although such glands are clearly distinguishable from adenomatous glands, they occasionally show adenomatous changes and the syndrome may rarely be accompanied by APs and carcinomas of the colon (Giardiello et al., 1987; see also Perzin & Bridge, 1982).

4. Polypoid Carcinoma

Polypoid carcinoma refers to a polyp consisting mainly or exclusively of malignant glands. Even when no adenomatous remnants can be identified in them, it is believed that many of these lesions originate in APs as "malignant polyps" and the expanding carcinoma has destroyed the residual adenoma. Although they frequently exhibit macroscopic stigmata of malignancy such as surface irregularity, induration, or ulceration, some of them may mimic benign APs.

5. Nonepithelial Polyps

The most common of the nonepithelial polyps consist of *mucosal tags*, which are protrusions of edematous or thickened mucosa that may show elongated crypts histologically, and *inflammatory polyps*. The latter, also known as *pseudopolyps*, show inflammatory infiltrates and granulation tissue admixed with nonneoplastic colonic glands. They are typically found in inflammatory bowel disease, especially ulcerative colitis, and may be grossly indistinguishable from APs, but in inflammatory bowel disease, the presence of adjacent mucosal ulcerations generally suggests the correct diagnosis. Connective tissue tumors originating in the submucosa or muscularis propria, or lymphomas, may also present as polyps, but the presence of intact overlying mucosa usually permits recognition of their nonepithelial origin.

E. The Diminutive Polyp

In one often quoted study, 90 percent of tiny polyps were found to be HyPs (Lane et al., 1971). (One earlier investigation of 1,000 diminutive polyps did reveal 86 percent to be adenomatous [Pagtalunan et al., 1965], but no hyperplastic polyps were found, and it is difficult to evaluate their findings.) Being thus devoid of neoplastic potential, it was felt that such lesions did not have to be excised. However, the polyps examined by Lane et al. were < 3 mm, whereas diminutive polyps by definition are < 5 mm. More recent studies indicate that a substantial number of diminutive polyps are adenomatous, ranging from 37 to 79 percent (Granqvist et al., 1979; Tedesco et al., 1982; Feczko et al., 1984; Gottlieb et al., 1984; Isbister, 1986; Wegener et al., 1986). Several authors have emphasized that a significantly higher percentage of polyps in the 4–5 mm range are adenomatous than those between 1 and 3 mm. The vast majority of the lesions are well-differentiated tubular adenomas, but moderate to severe dysplasia and even invasive carcinoma are occasionally found in them.

F. Other Characteristics of Polyps

1. Kinetics

In contrast to normal mucosa, in which the ability to synthesize DNA is confined to the lower two thirds of the crypts, adenomatous epithelium retains this ability up to the surface epithelium (Lipkin et al., 1963). Evidence for this includes the presence of mitoses noted histologically and of surface cells labeled by DNA precursors, such as tritiated thymidine, in radioautographs. A similar retention of replicative capacity is found in colorectal cancer cells. Increases in labeling index (the number of labeled epithelial cells/total number of cells) are also described in APs and found to be proportional to the degree of dysplasia (Bleiberg et al., 1985; Kanemitsu et al., 1985). The expansion of the proliferative compartment to include surface epithelium is also found in morphologically normal epithelium in patients with familial polyposis and, along with increased labeling index, in some normal mucosae adjacent to polyps from nonpolyposis patients (Deschner & Lipkin, 1975).

It has been found that calcium, added to their diet or to the tissue culture medium, reduces colonic epithelial cell proliferation, as determined by tritiated thymidine labeling in cultured normal colonic biopsy specimens from subjects at risk for familial CRC (Buset et al., 1986). No such inhibition was found in adenomatous or carcinomatous tissue. An epidemiological study from Hawaii failed to find a correlation between dietary calcium levels and risk of colon carcinoma, however (Heilbrun et al., 1986). It has been postulated that the antiproliferative effects of calcium on nonneoplastic colonic mucosa result from binding by calcium to free fatty acids and unconjugated bile acids. These compounds may be toxic to the mucosa and could be responsible for the putative tumor promoter effects of dietary fats.

Information similar to that obtained with tritiated thymidine was found by Risio et al. (1988) using an immunohistochemical technique based on a monoclonal antibody directed against bromodeoxyuridine, a pyrimidine analogue of thymidine introduced into biopsy specimens during in vitro incubation. The technique is less time-consuming than radioautography and avoids the use of radioisotopes.

2. Electron Microscopy

Earlier work showed no major difference between adenomatous and normal epithelium (Imai et al., 1965). It did demonstrate gaps in the lamina densa of the epithelial basement membrane in villous adenomas, through which epithelial cytoplasm extended into the lamina propria, and other findings indicative of decreased cell cohesiveness. A subsequent study (Kaye et al., 1973) showed defective maturation of the adenomatous epithelium. The adenomatous cells of the upper crypts consisted mainly of incompletely differentiated absorptive and goblet cells, both of which are usually confined to the lower third of normal crypts. The absorptive cells, for example, had fewer and shorter microvilli than expected at the same crypt level, and absorptive and goblet cells had less rough endoplasmic reticulum. The same authors described nests of dysplastic cells within the epithelial layer suggestive of early carcinoma in situ, a finding that can also be seen with the light microscope (see Figure 2.8). Other workers (Mughal & Filipe, 1978) described a continuum between normal–adenoma–carcinoma, and emphasized electron-dense cytoplasmic bodies that increased with progressive dysplasia and were found in malignant cells. Although the bodies were also found in transitional (normal) mucosa adjacent to carcinomas, the authors felt that such bodies, believed to be glycoproteins, had a role in malignant transformation. Electron microscopic features such as long rootlets extending down from microvilli, supernumerary desmosomes, apical vesicles, and variable numbers of glycocalyceal bodies have been noted in both APs and carcinomas and are said to constitute a "colonic ultrastructural profile" (Seiler et al., 1984). (See Figure 2.23.)

Using scanning electron microscopy, APs were found to have a characteristic surface appearance distinguishing them from normal, hyperplastic, and malignant epithelium (Fenoglio et al., 1975; Phelps et al., 1979). Typical tubular adenomas (TAs) had flat surfaces with irregular clefts, and VAs had cerebriform folds. Abnormalities in the mucus layer adjacent to APs have also been recorded (Traynor et al., 1981), but it is not known if these represent primary or secondary changes.

3. Histochemistry

Histochemical studies were designed to reveal metabolic or biochemical information about adenomatous epithelium and its relationship to carcinoma.

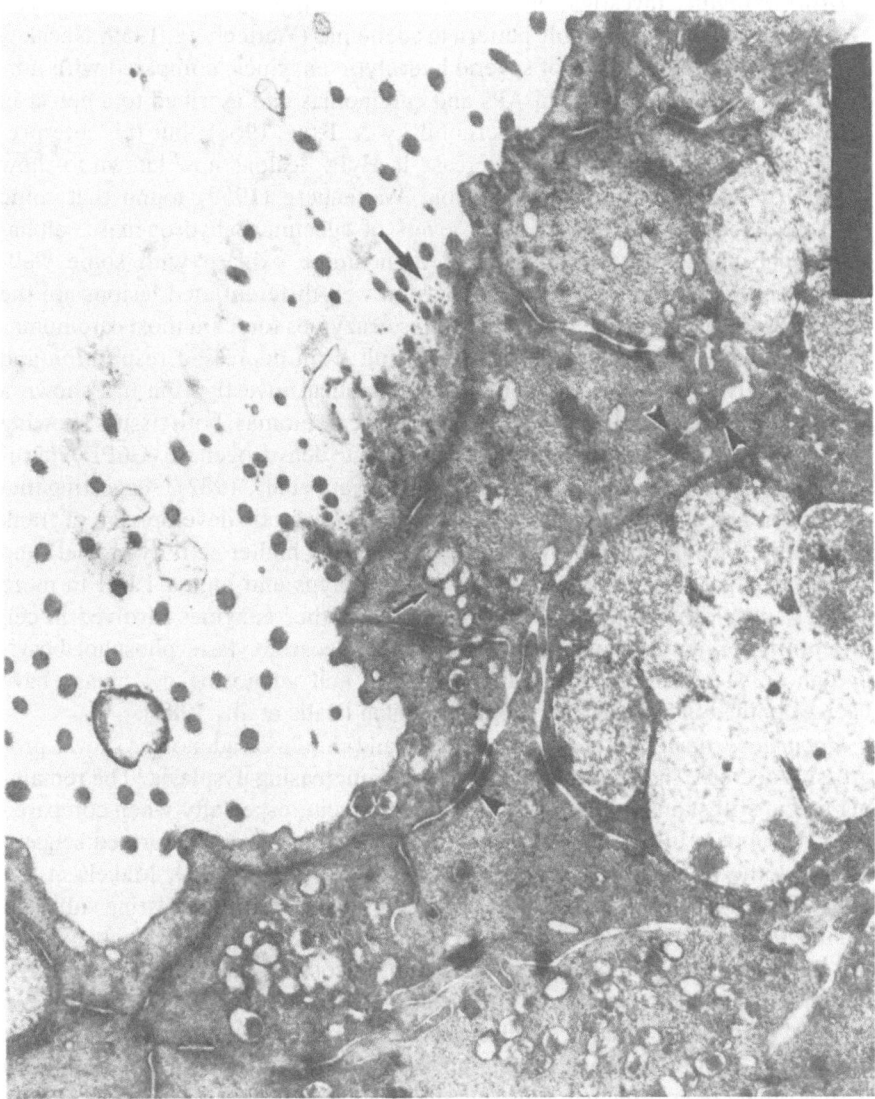

FIGURE 2.23. Apical portions of several absorptive and goblet cells in adenoma from an 8-year-old boy with Gardner's syndrome. They exhibit ultrastructural features frequently found in adenomas such as sparse microvilli, glycocalyceal bodies (long arrow), supernumerary desmosomes (arrowheads), and apical vesicles (short arrow). Electron micrograph; ×28,600, reproduced at 60%. (Courtesy of Dr. Shirley Siew.)

Histoenzymatic investigations of hydrolytic, respiratory, and proteolytic enzymes have shown a variable pattern in adenomas (Wattenberg, 1959; Nachlas, 1961). Decreased activities of several hydrolytic enzymes, compared with normal colon, have been found in APs and carcinomas and ascribed to a defect in maturation or differentiation (Czernobilsky & Tsou, 1968), but this interpretation is clouded by the similar decrease in HyPs, lesions now known to show normal or even "excessive" maturation. Wattenberg (1959) found that some dysplastic foci in APs shared high levels of succinic dehydrogenase, alpha-glycerophosphate dehydrogenase, and monoamine oxidase with some well-differentiated carcinomas. These changes in well-differentiated lesions are the reverse of the usual reduced activity of those enzymes found in most carcinomas in which the low levels are believed to result from depressed respiration and increased glycolysis. One more recent biochemical investigation has shown a more widespread congruence between APs and carcinomas, both tissues showing higher-than-normal levels of glucose-6-phosphate dehydrogenase (G6PD), pyruvate kinase, and lactic dehydrogenase (LDH) (Vatn et al., 1982), suggesting that changes in glycolysis or pentose metabolism may precede development of frank carcinoma. Further work from that group indicated higher activity of LDH and G6PD in larger (> 1.0 cm) than in smaller polyps and higher LDH in more severely dysplastic polyps (Hoff et al., 1985). Other enzymes involved in cell differentiation and turnover such as ornithine decarboxylase, phosphoribosyl-pyrophosphate synthetase, thymidine kinase, and adenosine deaminase have been examined in normal and neoplastic colon (Balis et al., 1984).

Mucin histochemistry studies of both adult and childhood adenomas show a progressive decrease in the amount of mucin with increasing dysplasia. The remaining mucin still stains fairly intensely for sulfomucin, especially when compared with sialomucin in the high iron diamine–alcian blue stain performed sequentially for the two acid mucins on the same slide (Filipe, 1969; Makela et al., 1971; Taki, 1980; Lev et al., 1987). The significance of this persisting sulfomucin in APs and its possible relationship to carcinogenesis, if any, are unknown. In fact, one dissenting group found no predominance of sulfomucin in APs compared with nonneoplastic polyps and no relationship between amount of sulfomucin in APs and their size, degree of dysplasia, or percentage of villous component (Grigioni et al., 1983). Another finding in APs is increased PAS activity compared with normal colon. This parallels the increase found in carcinomas and may be the result of reduced O-acylated sialomucins (Culling et al., 1977; P.E. Reid et al., 1980; Greaves et al., 1984; Szczepanski & Stachura, 1985; Lev et al., 1987). The abundance of that compound in normal colonic mucosa is believed to account for the weak PAS there (Culling et al., 1981), since the acyl groups are attached to the C8 hydroxyl groups of sialic acid rendering it resistant to periodic acid oxidation. Of related interest is the 82 percent loss of such O-acylated sialomucins in flat normal mucosa from patients with familial polyposis coli (Muto et al., 1985). This abnormal staining pattern was postulated to be a marker for future neoplasia in a morphologically normal mucosa, but the significance of these findings is unclear in view of the findings from the same group that even their normal colon control mucosae showed a 39 percent loss of those sialomucins. An alternate explanation for the increased PAS reactivity in APs is that this reflects the presence of a gastric fucomucin-associated antigen found in APs but not found in normal colonic mucosa (Bara et al., 1983); fucose is strongly PAS positive.

In both APs and carcinomas the general reduction of mucin may be accompanied by an increase in cytoplasmic glycogen (Tormo et al., 1987). Fetal colon also contains abundant glycogen, whereas normal adult colon is devoid of it.

4. Markers

Altered expression of lectins has been described in colonic neoplasia. Peanut agglutinin (PNA), which has an affinity for galactose $\beta1 \rightarrow 3$ N-acetylgalactosamine in oligosaccharide side chains of mucin molecules, is increased in carcinomas and some APs and shows a different cellular localization from normal (Boland et al., 1982; Cooper & Reuter, 1983; Yuan et al., 1986; Lev et al., 1987; Lance & Lev, submitted for publication). Some workers have found PNA binding to be stronger in larger and more severely dysplastic polyps, but others have not. Increased PNA staining, which may result from incomplete glycosylation and unmasking of T blood group antigens (precursors of MN antigens), is characteristic of colonic and other neoplasms (Springer, 1984).

Staining with *Ulex europaeus* agglutinin (UEA), which detects terminal L-fucosyl groups, has been detected in distal APs and carcinomas in both familial polyposis coli (FPC) and in subjects with sporadic APs but is not found in normal distal (left) colon (Yonezawa et al., 1982, 1983). This suggests reexpression of a fucose-associated compound, possibly blood group H(0) or other fucosylated antigens of Lewis blood group type known to be increased in colorectal carcinoma and in some APs (Yuan et al., 1987). Grossly normal left colon mucosa from patients with FPC was found by Yonezawa et al. to stain with UEA, whereas left-sided mucosa from nonpolyposis patients did not. This constituted another indication to those workers that this mucosa in FPC is at risk for future neoplasia. However, Sugihara and Jass (1987) found no difference in UEA binding by left colon in FPC and controls. Reappearance of blood group substances A and B, which are normally found in distal colon only in fetal life, has also been described in distal APs (Denk et al., 1975; Cooper et al., 1980). Staining of APs in distal colon for *Dolichos biflorus* agglutinin (DBA) is generally about the same as or less than that found in normal mucosa (Lev et al., 1987); since DBA binds to N-acetylgalactosamine, the determinant for blood group A, this appears to contradict the above findings, but it must be remembered that N-acetylgalactosamine is undoubtedly found as a terminal residue on oligosaccharide side chains other than those in blood group A. Abnormalities in staining of APs for other lectins such as *Ricinus communis* agglutinin and *Griffonia simplicifolia II* have also been described (Klein et al., 1981; Higgins et al., 1983; Rhodes et al., 1986).

Numerous investigators have demonstrated increased expressivity of carcinoembryonic antigen (CEA) in APs, especially those with a significant villous component or with severe dysplasia, and in carcinomas (see O'Brien et al., 1981;

Jothy et al., 1988). A progressive increase in CEA staining was seen in the same villous adenoma biopsied repeatedly over a six-year period prior to its transformation into invasive carcinoma (Whitehead & Skinner, 1983). Larger APs are often accompanied by elevated CEA in colonic lavage fluid and less commonly in plasma (Doos et al., 1975; Winawer et al., 1977). In addition to increased expression in neoplasms, CEA also shifts from an apical surface localization in normal colonic epithelium to the entire periphery of cancer cells. The apical localization is apparently retained in APs, suggesting to some workers that loss of CEA surface polarity is a late stage in carcinogenesis (Ahnen et al., 1987).

A shift has also been noted in the location of villin, a cytoskeletal protein normally confined to the microvillous border of colonic epithelial cells. Both APs and CRCs show reactivity for villin in such ectopic sites as the lateral and basement membrane areas and the cytoplasm, in addition to the usual apical localization (West et al., 1988). Since cytoplasmic localization is also found in fetal colon, at least in the chick, villin can be regarded as an oncofetal antigen, thus resembling CEA. A variety of other tumor-associated antigens have been found in APs (Skinner & Whitehead, 1981; Zotter et al., 1987). Per contra there has been a loss in APs of such markers of differentiation as IgA and secretory component (Rognum et al., 1982) and certain endocrine peptides (Adrian et al., 1988).

An observation of potential clinical interest concerns the association between skin tags (acrochordons) and APs. Piette et al. (1988) reviewed the literature and found a statistically significant association between the two lesions in subjects with gastrointestinal symptoms but not in asymptomatic individuals. They confirmed the association in the former group in their own prospective study of 100 subjects in which the colonoscopists were "blinded" to the skin findings. They felt that since people with gastrointestinal symptoms would in any case have been subjected to colonoscopy, the detection of skin tags in them has no practical value.

Although this review has concentrated on lectins/blood group antigens, CEA, and a cytoskeletal protein, the term *biomarker* can embrace other biochemical changes within colorectal neoplasms or their precursors, such as those involving DNA kinetics, glycoproteins, or enzymes, as discussed elsewhere in this chapter. One may ponder the significance of these markers. They may be related to tumorigenesis and provide information on the genesis of APs from normal mucosa or of CRC from APs or other precursors, or they may represent epiphenomena. Some have clinical significance: Increasing serum CEA may presage early recurrence of carcinoma, for example. Other markers—for example, uptake of labeled precursors of DNA by normal colonic surface epithelium—may define tissues or populations at risk for future neoplasia. It is possible that in the future some of these or analagous markers may help to identify early and treatable lesions, such as APs, foci of severe dysplasia in ulcerative colitis, or low-stage carcinoma. Finally, markers may in the future help determine those subsets of that very large population of subjects with known APs who are at especially high risk for subsequent carcinoma, thereby conserving surveillance resources and reducing costs.

5. Tissue Culture

APs have characteristics intermediate between normal and malignant mucosa. Growth of tubular adenomas, but not of tubulovillous or villous adenomas, is stimulated by epidermal growth factor (Friedman et al., 1981). This may be because tubular adenomas are the closest of the three to normal colonic cells, and it is known that normal cells are more responsive than transformed cells to exogenous epidermal growth factor (Shields, 1978). The abnormalities that APs do show are generally more numerous when such risk factors for malignancy as significant dysplasia or villous components are present. Chromosomal aberrations are more common in villous than in tubular adenomas, for example. When tissue culture preparations are stimulated with the tumor promoter TPA (a phorbol ester), APs with significant dysplasia or villosity, and carcinomas, secrete the protease plasminogen activator (plasmin), whereas well-differentiated tubular adenomas and normal mucosa do not (Friedman, 1985). She speculated that that enzyme might help the carcinoma invade and destroy adjacent adenomatous tissue and thus account for the frequent inability to find adenomatous remnants in carcinomas. Polyp cells, unlike carcinomas, are not tumorigenic in nude mice and do not show anchorage-independent growth (Paraskeva et al., 1984; Willson et al., 1987). In another study, normal colonic cells never survived when transplanted into nude mice, whereas adenoma cells survived for 28 days and colon carcinoma cells for 43 days (Bhargava & Lipkin, 1981). One investigator found that APs maintained for 3–14 days in a fibrin foam matrix showed histological progression into in situ and then invasive carcinoma (Kalus, 1972). This observation provides strong support for the adenoma–carcinoma sequence but unfortunately has not been described in subsequent reports by others.

6. DNA Content

In earlier years, nuclear DNA was examined by using microspectrophotometry or by utilizing stained chromosome spreads (Stich et al., 1960; Enterline & Arvan, 1967). The latter group demonstrated chromosomal abnormalities consistently in APs: Hyperploidy and structurally abnormal individual chromosomes were more common in the more severely dysplastic or villous tumors. More recently, flow cytometry has been the preferred method of study. Homogenates of nuclei obtained from neoplastic and normal colon are stained and the histograms subsequently generated yield information on DNA ploidy at various stages of the cell cycle. Traditionally, fresh specimens have been the source of the nuclei, but recently paraffin-embedded material has been used. DNA staining intensity and the percentage of detectable aneuploidy are said to be less with paraffin, however (Frierson, 1988). Initial work with colon carcinomas indicated that advanced Dukes stages had higher percentages of aneuploidy than lower stages and that tumors with aneuploidy were associated with a poorer prognosis than were diploid tumors even when stage matched (Wolley et al., 1982; Kokal et al., 1986).

APs have a lower percentage of aneuploidy than carcinomas; in one study (Goh & Jass, 1986), aneuploidy was found in 13 percent of 269 APs and in 64 percent of 203 carcinomas. The incidence of aneuploidy rose with the increase in dysplasia and villous component (Goh & Jass, 1986; Banner et al., 1987) and was higher in APs from patients with a family history of CRC than in those without such a history (Sciallero et al., 1987). Similar nuclear DNA content and nuclear size were found in foci of severe dysplasia and carcinoma in the same polyp (Jarvis et al., 1987). Some groups, however, found no correlation between cell cycle fractions and the severity of dysplasia or polyp size (Hoff et al., 1985), or between aneuploidy and the presence of villosity in APs (Weiss et al., 1985), but these are exceptions to the generally good correlation between degree of aneuploidy (and other chromosomal and cell cycle abnormalities) and risk factors for carcinoma in APs.

It has been found that crater-shaped invasive CRCs show greater DNA aneuploidy than do severely dysplastic or malignant polyps (Hamada et al., 1988). That group concluded that 60 percent of the crater-shaped carcinomas develop from nonpolypoid (? de novo) carcinomas but conceded that other factors could explain the difference in frequency of aneuploidy between the crater-shaped and polypoid lesions.

In situ hybridization techniques have been recently applied to colon tumors. In one such study (Mariani-Costantini et al., 1989), c-myc mRNA was found in low levels in the proliferative zone of normal mucosa and in progressively increasing amounts in villous adenomas and carcinomas respectively.

7. Doubling Time

Information on volumetric doubling time of colonic tumors stems largely from radiological surveys conducted in the precolonoscopic era when many polypoid lesions were observed over prolonged periods, since it was felt that the risk of operative removal exceeded the risk of malignancy. Net tumor growth of course depends not only on formation of new cells but also on cell death, exfoliation of tumor cells, and metastatic dissemination. Early tumor growth is geometric, whereas it is slower in larger tumors (Gompertzian). It was found that the doubling times of APs as a group were longer than those of carcinomas, one generally quoted median figure for carcinomas being 20 months (Welin et al., 1963). In one study (Figiel et al., 1965), it was found that 285 of 300 presumably benign polyps, most of which were < 1.5 cm in size, grew very slowly or not at all during the two- to nine-year follow-up period. There are very few data on the exact doubling times of polyps that are subsequently determined histologically to be benign, however. In one such study, 6 of 11 APs were found to have doubling times of 147 to 399 days, which overlapped with those of 6 carcinomas that ranged from 92 to 1,032 days, whereas 5 other APs did not grow at all during the 2.5-year period of the study (Tada et al., 1984). Table 2.3 summarizes a recent review by Spratt and Spratt (1985) that shows doubling times from four earlier studies, including the Welin work cited above. These results have implications for

TABLE 2.3. Mean doubling times for adenomatous polyps, carcinomas, and metastases.

	n	Arithmetic mean (days)	Range (days)
Polyps	77	1,600	13–8,664
Carcinomas	98	668	52–10,000
Metastases	23	221	32–3,300

SOURCE: Modified from Spratt and Spratt, 1985.

decisions on whether or not to perform polypectomy for diminutive polyps (Chaps. 2.E and 5.B).

8. Regression of Polyps

Occasional patients with familial polyposis exhibit regression of rectal adenomas or absence of new lesions after total colectomy and ileoproctostomy (Cole & Holden, 1959). In a larger series of patients, Bussey (1975) has confirmed this, especially in younger individuals. Possible reasons for this phenomenon include diversion of ileal contents to the rectum, loss of diseased colon, and rectal ischemia. Regression of sporadic polyps has also been described (Knoernschild, 1963). In a study of diminutive polyps left in situ for two years prior to excision, 5 of 35 sporadic APs showed reduction in mean size from 3.6 ($\pm$ 0.2) to 2.4 mm ($\pm$ 0.2 mm) ($p = 0.05$), and other polyps disappeared completely (Hoff et al., 1986). The latter polyps were unfortunately of unknown histology, since they were not biopsied initially, but it may be assumed that at least some of them were adenomatous.

G. Histogenesis of Polyps

Possible etiological factors in polyp formation such as genetic effects and luminal influences will be discussed in Chaps. 3.C and 4.D, respectively.

There is disagreement as to where in the crypts the initial changes originate. In their study of tiny APs in FPC, Lane and Lev (1963) demonstrated the gradual development of dysplasia in the deep crypt cells as they migrated upward and suggested that the adenomatous epithelium then spread laterally, imparting a "vase of flowers" appearance to the polyp. Upward progression of low- to higher-grade dysplasia within longitudinally sectioned crypts has been demonstrated not only histologically but histochemically (Lev et al., 1987). In sections of the polyp not taken through its center, the surface adenomatous changes overlying normal crypts may erroneously suggest a superficial origin of the adenomatous epithelium; this conclusion has been reached by several investigators who have also noted the shift in APs of the proliferative compartment from the lower to the upper crypt area (Cole & McKalen, 1963; Lipkin, 1974; Wiebecke et al., 1974; Maskens, 1979). However, further evidence of the deep crypt origin of adenoma formation is provided by histochemical comparisons between adenomatous and adjacent normal epithelium. For example, there is a predominance of sulfomucin over sialomucin in adenomatous epithelium, which is similar to the sulfomucin predominance in normal deep crypts. In addition, DBA, a lectin with an affinity for N-acetylgalactosamine, stains both adenomatous and normal deep crypt goblet cells weakly but stains normal upper crypt goblet cells much more intensely (Lev et al., 1987). Staining for Le^y, a blood group antigen, was found throughout many adenomas, whereas it was confined to the deep crypts in normal colon (Ruggiero et al., 1988). Of course, histochemical similarities do not necessarily reflect common ancestry, but the evidence is supportive of such a relationship. Oohara et al. (1982) also suggested that adenomas arose from basal cells in deep crypts. They stated that 100 percent of APs in nonpolyposis cases and 82 percent of APs in familial polyposis arose in that fashion. In the remaining familial polyposis cases, the fact that adenomas did not originate from deep basal cells could be explained by the exclusive presence of mature goblet cells in the deep crypts from which one would not expect adenomatous epithelium to arise. In such cases, they speculated, the adenomas might have arisen from less differentiated cells in the mid- or upper mucosa.

In other studies of FPC, a fertile source of investigation of the genesis of early lesions, adenomatous change has been detected in a single crypt or portion thereof (Bussey, 1975; Oohara et al., 1982; Nakamura & Kino, 1984). Minute adenomatous foci involving one to two crypts have also been found in grossly normal mucosa from nonpolyposis patients (Woda et al., 1977; Lev & Grover, 1981).

Another possible precursor of adenomatous epithelium is the eosinophilic epithelium described by Urbanski et al. (1986). In that condition, the crypts contain reduced numbers of goblet cells and are lined by cells with strongly eosinophilic cytoplasm. That group found frequent transitions between, or juxtapositions of, eosinophilic and adenomatous epithelium. Eosinophilic crypts have been noted in the absence of adenomatous glands, however (Lev & Grover, 1981). The possible role of the lymphoid follicles, which may be found beneath adenomatous epithelium, in the genesis of APs in humans and experimental animals has been discussed by Oohara et al. (1981) and by others (see Deasy et al., 1983). Oohara's illustrations of glands adjacent to such follicles are not convincingly adenomatous, however, and it is known that basophilic epithelium of nonadenomatous type may be found in that location (see Chapter 1).

The proliferating adenoma cells are believed to represent a failure of differentiation of the precursors of absorptive and goblet cells as documented in ultrastructural and kinetic studies (see Chap. 2.F). It is uncertain if the origin of adenoma is monoclonal or polyclonal. Using G6PD analyses of adenoma cells, evidence has been obtained for both pathways (see Haggitt & Reid, 1986). Monoclonality was found in a recent study which excluded contaminating stroma from the DNA analysis (Fearon et al., 1987).

A stereological analysis of polyp growth in 14 APs (presumably TAs) from a patient with FPC has been performed by Pesce and Colacino (1987), who calculated the area of glandular and surface epithelium. They found that glandular expansion accounted for most of the increase in polyp volume.

The relationship between the villous and tubular components of APs has occasioned some controversy. Some workers believe that there is a continuum between these two forms of AP (Bussey, 1975; Fenoglio et al., 1977; Christie, 1981; also see Enterline, 1976). Villous structures are often found on the surface of polyps and tubules below, as noted by Potet and Soullard (1971) and illustrated in Figure 2.6, implying transformation of tubules into villi. The vast majority of diminutive polyps are tubular, whereas larger polyps more frequently contain villi, suggesting that as polyps grow, they acquire more villous characteristics. The opposing school believes that the two types of adenomas are different from their inception and that transition forms do not exist. They point out that very small exclusively villous polyps do exist, as do large and strictly tubular adenomas, and feel that different mechanisms account for the two patterns. Earlier workers (see Grinnell & Lane, 1958) suggested that the growth stimulus in VA acts mainly on the surface cells, whereas in TA it acts on mid- or deep crypts. More recently, it has been hypothesized that the nonproliferative stroma of the tubular adenoma "impedes" the growth of the adenomatous tubules, which then undergo infoldings, whereas in villous adenomas the stromal growth accompanies the epithelial proliferation, resulting in an outward (upward) expansion (Wiebecke et al., 1974; Maskens, 1979). Wiebecke et al. (1974) suggested that villous adenomas resulted from vertical "foliaceous" growth with groovelike lengthening of crypts. In their stereological study, Elias et al. (1981) stated that "villi" represented sections of what were really branching folia and that these folia resulted not from proliferating lamina propria, as suggested by Wiebecke et al. (1974), but from progressive fusion of adjacent crypt epithelial layers and of adjacent lumina in the upper crypts.

Irrespective of where in the crypt the initial adenomatous change starts, the main proliferation of adenomatous tissue is indeed toward the luminal surface, resulting in the typical convex upper surface of the classical AP. Although all polyps start as sessile growths, some of them then become pedunculated, presumably as the result of intestinal peristalsis and other factors such as the quality of the feces and the presence of a mesocolon. It has been pointed out that in the descending and sigmoid colon, where there is strong peristalsis and well-formed stools, polyps tend to be pedunculated more often than in colonic segments in which such features are not present (Christie, 1981).

The transformation of AP into invasive carcinoma has been the subject of several investigations. A progressive increase in plasminogen activator activity has been noted from normal to adenomatous to malignant epithelium by de Bruin et al. (1987); Gelister et al. (1987) felt that plasmin generated from plasminogen activator could play a role in conversion of AP into carcinoma by digesting the epithelial basement membranes of polyps. Forster et al. (1986) have noted a loss in laminin from the basement membrane in some rectal carcinomas, a feature associated with a poorer prognosis. Malignant glands invading the lamina propria or submucosa generally invoke a fibroblastic response not found with in situ carcinomas confined by a basement membrane. This may be the result of a growth factor produced by tumor cells and/or represent a component of the "host response" to the tumor.

References

Adrian TE, Ballantyne GH, Zucker KA, Zdon MJ, Tierney R, Modlin IM (1988) Lack of peptide YY immunoreactivity in adenomatous colonic polyps: Evidence in favor of an adenoma–carcinoma sequence. J Surg Res 44:561–565.

Ahnen DJ, Kinoshita K, Nakane P, Brown WR (1987) Differential expression of carcinoembryonic antigen and secretory component during colonic epithelial cell differentiation and in colonic carcinomas. Gastroenterology 93:1330–1338.

Arthur JF (1968) Structure and significance of metaplastic nodules in the rectal mucosa. J Clin Pathol 21:735–743.

Balis ME, Yip LC, Rinaldy A, Relyea NM, Higgins PJ, Kemeny NE (1984) Enzymic and metabolic markers of colon cancer. In SR Wolman, AJ Mastromarino (eds): Markers of Colonic Cell Differentiation. Raven Press, New York, pp 295–308.

Banner BF, Chacho MS, Roseman DL, Coon JS (1987) Multiparameter flow cytometric analysis of colon polyps. Am J Clin Pathol 87:313–318.

Bara J, Languille O, Gendron MC, Daher N, Martin E, Burtin P (1983) Immunohistochemical study of precancerous mucus modification in human distal colonic polyps. Cancer Res 43:3885–3891.

Bengoechea O, Martinez-Penuela JM, Larrinaga B, Valerdi J, Borda F (1987) Hyperplastic polyposis of the colorectum and adenocarcinoma in a 24-year-old man. Am J Surg Pathol 11:323–327.

Berg JW (1988) Epidemiology, pathology and the importance of adenomas. In G Steele, RW Burt, SJ Winawer, JP Karr (eds): Basic and Clinical Perspectives of Colorectal Polyps and Cancer. Alan R Liss, New York, pp 13–21.

Bhargava DK, Lipkin M (1981) Transplantation of adenomatous polyps, normal colonic mucosa and adenocarcinoma of colon into athymic mice. Digestion 21:225–231.

Bleiberg H, Buyse M, Galand P (1985) Cell kinetic indicators of premalignant stages of colorectal cancer. Cancer 56:124–129.

Boland CR, Montgomery CK, Kim YS (1982) Alterations in human colonic mucin occurring with cellular differentiation and malignant transformation. Proc Natl Acad Sci USA 79:2051–2055.

Brown LJR, Smeeton NC, Dixon MF (1985) Assessment of dysplasia in colorectal adenomas: An observer variation and morphometric study. J Clin Pathol 38:174–179.

Buset M, Lipkin M, Winawer S, Swaroop S, Friedman E (1986) Calcium, cellular proliferation and cancer. Cancer Res 46:5426–5430.

Bussey H (1975) Familial Polyposis Coli. Johns Hopkins University Press, Baltimore.

Chapuis P, Dent O, Goulston K (1982) Clinical accuracy in the diagnosis of small polyps using the flexible fiberoptic sigmoidoscope. Dis Colon Rectum 25:669–672.

Christie J (1981) Comparative significance of right colon, left colon and rectal polyps. Gastrointest Endosc 27(1):185–186.

Clark JC, Collan Y, Eide TN, Esteve J, Ewen S, Gibbs NM, Jensen OM (1985) Prevalence of polyps in an autopsy series from areas with varying incidence of large-bowel cancer. Int J Cancer 36:179–186.

Cole JW, Holden WD (1959) Postcolectomy regression of adenomatous polyps of the rectum. AMA Arch Surg 79:385–392.

Fenoglio CM, Kaye GI, Pascal RR, Lane N (1977) Defining the precursor tissue of ordinary large bowel carcinoma: Implications for cancer prevention. Pathol Annu 12:87–116.

Fenoglio CM, Richart RM, Kaye GI (1975) Comparative electron microscopic features of normal, hyperplastic and adenomatous human colonic epithelium: Variations in surface architecture found by scanning electron microscopy. Gastroenterology 69:100–109.

Fenoglio-Preiser CM (1985) Polyps and the subsequent development of carcinoma of the colon and rectum: Definitions and hints on tissue handling. In CM Fenoglio-Preiser, FP Rossini (eds): Adenomas and Adenomas Containing Carcinoma of the Large Bowel: Advances in Diagnosis and Therapy. Raven Press, New York, pp 15–29.

Feyrter F (1929) Zur Lehre von der Polypenbildung im menschlichen Darm. Wiener Med Wochensch 79:338–342.

Figiel LS, Figiel SJ, Wietersen FK (1965) Roentgenologic observations of growth rates of colonic polyps and carcinoma. Acta Radiol [Diagn] (Stockh) 3:417–429.

Filipe MI (1969) Value of histochemical reactions for mucosubstances in the diagnosis of certain pathological conditions of the colon and rectum. Gut 10:577–586.

Forster SJ, Talbot IC, Clayton DG, Critchley DR (1986) Tumor basement membrane laminin in adenocarcinoma of rectum: An immunohistochemical study of biological and clinical significance. Int J Cancer 37:813–817.

Friedman EA (1985) A multistage model for human colon carcinoma development from tissue culture studies. In JRF Ingall, AJ Mastromarino (eds): Carcinoma of the Large Bowel and Its Precursors. Alan R. Liss, New York, pp 175–186.

Friedman EA, Higgins P, Lipkin M, Shinya H, Gelb AM (1981) Tissue culture of human epithelial cells from benign colonic tumors. In Vitro 17:632–644.

Frierson HF (1988) Flow cytometric analysis of ploidy in solid neoplasms. Hum Pathol 19:290–294.

Frühmorgen P, Matek W (1983) Significance of polypectomy in the large bowel — endoscopy. Endoscopy 15:155–157.

Gelister JS, Lewin MR, Driver HE, Savage F, Mahmoud M, Gaffney PJ, Boulos PB (1987) Plasminogen activators in experimental colorectal neoplasia: A role in the adenoma–carcinoma sequence? Gut 28:816–821.

Giardiello FM, Welsh SB, Hamilton SR, Offerhaus GJA, Gittelsohn AM, Booker SV, Krush AJ, Yardley JH, Luk GD (1987) Increased risk of cancer in the Peutz-Jeghers syndrome. N Engl J Med 316:1511–1514.

Gillespie PE, Chambers TJ, Chan KW, Doronzo F, Morson BC, Williams CB (1979) Colonic adenomas — a colonoscopic survey. Gut 20:240–245.

Goh HS, Jass JR (1986) DNA content and the adenoma–carcinoma sequence in the colorectum. J Clin Pathol 39:387–392.

Goldman H, Ming SC, Hickok DF (1970) Nature and significance of hyperplastic polyps of the human colon. Arch Pathol (Chicago) 89:349–354.

Goodman ZD, Yardley JH, Milligan FD (1979) Pathogenesis of colonic polyps in multiple juvenile polyposis. Cancer 43:1906–1913.

Gottlieb LS, Winawer SJ, Sternberg S, Magrath C, Diaz B, Zauber A, O'Brien M (1984) National Polyp Study (NPS): The diminutive colonic polyp. Gastrointest Endosc 30:143. (Abstract)

Graham AR, Paplanus SH, Bartels PH (1988) Micromorphometry of colonic lesions. Lab Invest 59:397–402.

Granqvist S (1981) Distribution of polyps in the large bowel in relation to age. A colonoscopic study. Scand J Gastroenterol 16:1025–1031.

Granqvist S, Gabrielson N, Sunderlin BP (1979) Diminutive colonic polyps – clinical significance and management. Endoscopy 11:36–42.

Greaves P, Filipe MI, Abbas S, Ormerod MG (1984) Sialomucins and carcinoembryonic antigen in the evolution of colorectal cancer. Histopathology 8:825–834.

Greene FL (1983) Distribution of colorectal neoplasms: A left to right shift of polyps and cancer. Am J Surg 49:62–65.

Grigioni W, Miglioli M, Santini D, Vanzo M, Piccaluga A, Barbara L (1983) Histochemistry of mucins and carcinogenic sequence: A study of some gastric and colorectal lesions. In P Sherlock, BC Morson, L Barbara, U Veronesi (eds): Precancerous Lesions of the Gastrointestinal Tract. Raven Press, New York, pp 127–135.

Grinnell RS, Lane N (1958) Benign and malignant adenomatous polyps and papillary adenomas of the colon and rectum. An analysis of 1,856 tumors in 1,335 patients. Int Abst Surg 106:519–538.

Haggitt RC, Reid BJ (1986) Hereditary gastrointestinal polyposis syndromes. Am J Surg Pathol 10:871–887.

Hamada S, Namura K, Fujita S (1988) The possibility of nonpolypoid carcinogenesis in the large intestine as inferred from frequencies of DNA aneuploidy of polypoid and crater-shaped carcinomas. Cancer 62:1503–1510.

Hayashi T, Yatani R, Apostol J, Stemmermann GN (1974) Pathogenesis of hyperplastic polyps of the colon: A hypothesis based on ultrastructure and in vitro cell kinetics. Gastroenterology 66:347–356.

Heilbrun LK, Hankin JH, Nomura AMY, Stemmermann GN (1986) Colon cancer and dietary fat, phosphorus and calcium in Hawaiian-Japanese men. Am J Clin Nutr 43:306–309.

Hermanek P (1985) Diagnosis and therapy of cancerous adenoma of the large bowel: A German experience. In CM Fenoglio-Preiser, FP Rossini (eds): Adenomas and Adenomas Containing Carcinoma of the Large Bowel: Advances in Diagnosis and Therapy. Raven Press, New York, pp 57–62.

Hermanek P, Frühmorgen P, Guggenmoos-Holzmann I, Altendorf A, Matek M (1983) The malignant potential of colorectal polyps – a new statistical approach. Endoscopy 15:16–20.

Higgins PJ, Friedman E, Lipkin M, Hertz R, Attiyeh F, Stonehill EH (1983) Expression of gastric-associated antigens by human premalignant and malignant colonic epithelial cells. Oncology 40:26–30.

Hoff G, Clausen OP, Fjordvang H, Norheim A, Foerster A, Vatn MH (1985) Epidemiology of polyps in the rectum and sigmoid colon. Scand J Gastroenterol 20:983–989.

Hoff G, Foerster A, Vatn MH, Sauar J, Larsen S (1986) Epidemiology of polyps in the rectum and colon. Recovery and evaluation of unresected polyps two years after detection. Scand J Gastroenterol 21:853–862.

Imai H, Saito S, Stein A (1965) Ultrastructure of adenomatous polyps and villous adenomas of the large intestine. Gastroenterology 48:188–197.

Isbister WH (1986) Colorectal polyps: An endoscopic experience. Aust NZ J Surg 56:717–722.

Jarvinen H, Franssila KO (1984) Familial juvenile polyposis coli; increased risk of colorectal cancer. Gut 25:792–800.

Jarvis LR, Graff PS, Whitehead R (1987) Correlation of nuclear ploidy with histology in adenomatous polyps of colon. J Clin Pathol 40:26–33.

Jothy S, Boucher L, Fuks A (1988) Differential expression of CEA epitopes in colonic polyps correlates with histopathological features. Lab Invest 58:44A.

Kalus M (1972) Carcinoma and adenomatous polyps of the colon and rectum in biopsy and organ tissue culture. Cancer 30:972–982.

Kanemitsu T, Koike A, Yamamoto S (1985) Study of the cell proliferation kinetics in ulcerative colitis, adenomatous polyps and cancer. Cancer 56:1094–1098.

Kaye GI, Pascal RR, Lane N (1971) The colonic pericryptal fibroblast sheath: Replication, migration, and cytodifferentiation of a mesenchymal cell system in adult tissue. III. Replication and differentiation in human hyperplastic and adenomatous polyps. Gastroenterology 60:515–536.

Kaye G, Fenoglio C, Pascal R, Lane N (1973) Comparative electron microscopic features of normal, hyperplastic and adenomatous human colonic epithelium. Variations in cellular structure relative to the process of epithelial differentiation. Gastroenterology 64:926–945.

Klein PJ, Osmers M, Vierbuchen M, Ortmann M, Kania J, Uhlenbruck G (1981) The importance of lectin binding sites and carcinoembryonic antigen with regard to normal, hyperplastic, adenomatous and carcinomatous colonic mucosa. Recent Results Cancer Res 79:1–9.

Knoernschild HE (1963) Growth rate and malignant potential of colonic polyps: Early results. Surg Forum 14:137–138.

Kokal W, Sheibani K, Terz J (1986) Tumor DNA content in the prognosis of colorectal carcinoma. JAMA 255:3123–3127.

Konishi F, Morson B (1982) Pathology of colorectal adenomas: A colonoscopic survey. J Clin Pathol 35:830–841.

Kozuka S (1975) Premalignancy of the mucosal polyp in the large intestine. I. Histologic gradation of the polyp on the basis of epithelial pseudostratification and glandular branching. Dis Colon Rectum 18:483–493.

Lance P, Lev R (1989) Colonic oligosaccharide structure deduced from lectin-binding studies before and after desialylation. (Submitted for publication)

Lane N, Kaplan H, Pascal R (1971) Minute adenomatous and hyperplastic polyps of the colon: Divergent patterns of epithelial growth with specific associated mesenchymal changes. Gastroenterology 60:537–551.

Lane N, Lev R (1963) Observations on the origin of adenomatous epithelium of the colon. Serial section studies of minute polyps in familial polyposis. Cancer 16:751–764.

Lev R, Grover R (1981) Precursors of human colon carcinoma: A serial section study of colectomy specimens. Cancer 47:2007–2015.

Lev R, Lebenthal E, Rossi T, Lance P (1987) Histochemical and morphological analysis of colonic epithelium from children with Gardner's syndrome and adults bearing adenomatous polyps. J Pediatr Gastroenterol Nutr 6:414–425.

Lipkin M (1974) Phase I and phase II proliferative lesions of colonic epithelial cells in diseases leading to colonic cancer. Cancer 34:878–888.

Lipkin M, Bell B, Sherlock P (1963) Cell proliferation kinetics in the gastrointestinal tract of man. I. Cell renewal in colon and rectum. J Clin Invest 42:767–776.

Makela V, Korhonen LK, Lilus G (1971) Carbohydrate-rich compounds in the colonic mucosa of man. Cancer 27:120–127.

Mariani-Costantini R, Theillet C, Hutzell P, Merlo G, Schlom J, Callahan R (1989) In situ detection of c-myc mRNA in adenocarcinomas, adenomas, and mucosa of human colon. J Histochem Cytochem 37:293–298.

Maskens AP (1979) Histogenesis of adenomatous polyps in the human large intestine. Gastroenterology 77:1245–1251.

Morson BC, Bussey HJR, Day DW, Hill MJ (1983) Adenomas of large bowel. Cancer Surv 2:451–478.

Morson BC, Sobin L (1976) Histological Typing of Intestinal Tumors. World Health Organization, Geneva, pp 13–58.

Mughal S, Filipe MI (1978) Ultrastructural study of the normal mucosa–adenoma–cancer sequence in the development of familial polyposis coli. J Natl Cancer Inst 60:753–768.

Muto T, Bussey HJR, Morson B (1975) The evolution of cancer of the colon and rectum. Cancer 36:2251–2270.

Muto T, Kamiya J, Sawada T, Agawa S, Morioka Y, Utsunomiya J (1985) Mucin abnormality of colonic mucosa in patients with familial polyposis coli: A new tool for early detection of the carrier? Dis Colon Rectum 28:147–148.

Nachlas MN (1961) Histochemical observations on the polyp-carcinoma sequence. Surg Gynecol Obstet 112:534–542.

Nakamura S, Kino I (1984) Morphogenesis of minute adenomas in familial polyposis coli. J Natl Cancer Inst 73:41–49.

Nakayama H, Kondo Y, Saito N, Sarashina H, Okui K (1988) Morphometric analysis of cytological atypia in colonic adenomas. Virchows Arch [A] 413:499–504.

Neale AJ, Demers RY, Budey H, Scott RO (1987) Physician accuracy in diagnosing colorectal polyps. Dis Colon Rectum 20:247–250.

Nishizawa M, Okada T, Sato F, Kariya A, Mayama S, Nakamura K (1980) A clinicopathological study of minute polypoid lesions of the colon based on magnifying fibercolonoscopy and dissecting microscopy. Endoscopy 12:124–129.

O'Brien MJ, Zamcheck N, Burke B, Kirkham SE, Saravis CA, Gottlieb LS (1981) Immunocytochemical localization of carcinoembryonic antigen in benign and malignant colorectal tissues. Am J Clin Pathol 75:283–290.

Oohara T, Ogino A, Tohma H (1981) Microscopic adenoma in non-polyposis coli: Incidence and relation to basal cells and lymphoid follicles. Dis Colon Rectum 24:120–126.

Oohara T, Ihara O, Saji K, Tohma H (1982) Comparative study of familial polyposis coli and nonpolyposis coli on the histogenesis of large-intestinal adenoma. Dis Colon Rectum 25:446–453.

Pagtalunan RJG, Dockerty MB, Jackman RJ, Anderson MJ (1965) The histopathology of diminutive polyps of the large intestine. Surg Gynecol Obstet 120:1259–1265.

Paraskeva C, Buckle B, Sheer D, Wigley C (1984) The isolation and characterization of colorectal epithelial cell lines at different stages in malignant transformation from familial polyposis coli patients. Int J Cancer 34:49–56.

Perzin KH, Bridge MF (1982) Adenomatous and carcinomatous changes in hamartomatous polyps of the small intestine (Peutz-Jeghers syndrome). Cancer 49:971–983.

Pesce CM, Colacino R (1987) Relative growth of adenomatous polyps of the colon. Stereology and allometry of multiple polyposis. Virchows Arch [A] 412:151–154.

Phelps PC, Toker C, Trump BF (1979) Surface ultrastructure of normal, adenomatous, and malignant epithelium from human colon. In Scanning Electron Microscopy III. SEM Inc., AMF O'Hare, Ill., pp 169–179.

Piette AM, Meduri B, Fritsch J, Fermanian J, Piette JC, Chapman A (1988) Do skin tags constitute a marker for colonic polyps? Gastroenterology 95:1127–1129.

Potet F, Soullard J (1971) Polyps of the rectum and colon. Gut 12:468–482.

Reid BJ, Haggitt RC, Rubin CE, Roth G, Surawicz CM, Van Belle G, Lewin K, Weinstein WM, Antonioli DA, Goldman H, MacDonald W, Owen D (1988) Observer variation in the diagnosis of dysplasia in Barrett's esophagus. Hum Pathol 19:166–178.

Reid PE, Culling CFA, Dunn WL, Ramey CW, Magil AB, Clay MG (1980) Differences between the O-acetylated sialic acids of the epithelial mucins of human colonic tumors and normal controls: A correlative chemical and histochemical study. J Histochem Cytochem 28:217–222.

Rhodes JM, Black RR, Savage A (1986) Glycoprotein abnormalities in colonic carcino-mata, adenomata and hyperplastic polyps shown by lectin peroxidase histochemistry. J Clin Pathol 39:1331–1334.

Rickert RR, Auerbach O, Garfinkel L, Hammond EC, Frasca JM (1979) Adenomatous lesions of the large bowel. An autopsy survey. Cancer 43:1847–1857.

Rider JA, Kirsner JB, Moeller HC, Palmer WL (1954) Polyps of the colon and rectum. Their incidence and relationship to carcinoma. Am J Med 16:555–564.

Risio M, Coverlizza S, Ferrari A, Candelaresi GL, Rossini FP (1988) Immunohistochemi-cal study of epithelial cell proliferation in hyperplastic polyps, adenomas, and adeno-carcinomas of the large bowel. Gastroenterology 94:899–906.

Rognum TO, Fausa O, Brandtzaeg P (1982) Immunohistochemical evaluation of carcino-embryonic antigen, secretory component, and epithelial IgA in tubular and villous large-bowel adenomas with different grades of dysplasia. Scand J Gastroenterol 17: 341–348.

Roth SI, Helwig EB (1963) Juvenile polyps of the colon and rectum. Cancer 16:468–479.

Ruggiero F, Cooper HS, Steplewski Z (1988) Immunohistochemical study of colorectal adenomas with monoclonal antibodies against blood group antigens (sialosyl-Le^a, Le^a, Le^b, Le^x, Le^y, A, B, and H). Lab Invest 59:96–103.

Sarlin JG, Mori K (1984) Morules in epithelial tumors of the colon and rectum. Am J Surg Pathol 8:281–285.

Sciallero S, Bruno S, DiVinci A, Geido E, Aste H, Giaretti W (1988) Flow cytometric DNA ploidy in colorectal adenomas and family history of colorectal cancer. Cancer 61:114–120.

Seiler MW, Reilova-Velez J, Hickey W, Bono L (1984) Ultrastructural markers of large bowel cancer. In SR Wolman, AJ Mastromarino (eds): Markers of Colonic Cell Differentiation. Raven Press, New York, pp 51–67.

Shields R (1978) Growth factors. Nature 272:670–671.

Shinya H, Wolff WI (1979) Morphology, anatomic distribution, and cancer potential of colonic polyps. Ann Surg 190:679–683.

Skinner JM, Whitehead R (1981) Tumour-associated antigens in polyps and carcinomas of the human large bowel. Cancer 47:1241–1245.

Sobin LH (1985) Inverted hyperplastic polyps of the colon. Am J Surg Pathol 9:265–272.

Spjut HJ, Appel M (1979) Epithelial polyps of the large bowel: A pathological and colonoscopic study. Curr Probl Cancer 4:23–42.

Spjut HJ, Estrada RG (1977) The significance of epithelial polyps of the large bowel. Pathol Annu, Pt I, 12:147–170.

Spratt JS, Ackerman LV, Moyer CA (1958) Relationship of polyps of the colon to colonic cancer. Ann Surg 148:682–698.

Spratt JS, Spratt JA (1985) Growth rates of benign and malignant neoplasms of the colon. In JRF Ingall, AJ Mastromarino (eds): Carcinoma of the Large Bowel and Its Precur-sors. Alan R Liss, New York, pp 103–120.

Springer GF (1984) T and Tn, general carcinoma autoantigens. Science 224:1198–1206.

Stemmermann GN, Yatani R (1973) Diverticulosis and polyps of the large intestine. A necropsy study of Hawaii Japanese. Cancer 31:1260–1270.

Stich HF, Florian SF, Emson HE (1960) The DNA content of tumor cells. I. Polyps and adenocarcinomas of the large intestine of man. J Natl Cancer Inst 24:471–482.

Sugihara K, Jass JR (1987) Colorectal goblet cell mucins in familial adenomatous polypo-sis. J Clin Pathol 40:608–611.

Szczepanski W, Stachura J (1985) O-acylated sialomucins in adenomatous and hyperplas-tic colonic polyps. Mater Med Pol 17:13–15.

Tada M, Misaki F, Kawai K (1984) Growth rates of colorectal carcinoma and adenoma by roentgenologic follow-up observations. Gastroenterol Jpn 19:550–555.

Taki T (1980) Morphology and histochemistry of large intestinal polyps. Acta Pathol Jpn 30:355–363.

Tedesco FJ, Hendrix JC, Pickens CA, Brady PG, Mills LR (1982) Diminutive polyps: Histopathology, spatial distribution and clinical significance. Gastrointest Endosc 28:1–5.

Thompson JJ, Enterline HT (1981) The macroscopic appearance of colorectal polyps. Cancer 48:151–160.

Tormo B, Rodriguez T, Rengifo E, Mandado S, Gonzalez N, Gra B (1987) Ultrastructural study of glycogen containing cells in colonic adenocarcinomas and precancerous polypoid lesions. Arch Geschwulstforsch 57:39–46.

Traynor OJ, Costa NL, Blumgart LH, Wood CB (1981) A scanning electron microscopy study of the colonic mucosa and colonic mucus layer of patients with normal colons, adenomatous polyps and colorectal carcinoma. Br J Surg 68:701–704.

Urbanski SJ, Harber G, Hartwick W, Kortan P, Marcon N, Miceli P (1986) Mucosal changes associated with adenomatous colonic polyps. Am J Pathol 124:34–38.

Urbanski SJ, Marcon N, Kossakowska AE, Bruce WR (1984) Mixed hyperplastic adenomatous polyps—an underdiagnosed entity. Report of a case of adenocarcinoma arising within a mixed hyperplastic adenomatous polyp. Am J Surg Pathol 8:551–556.

Vatn MH, Tjora S, Arva PH, Serck-Hanssen A, Stromme JH (1982) Enzymatic characteristics of tubular adenomas and carcinomas of the large intestine. Gut 23:194–197.

Wattenberg LW (1959) A histochemical study of five oxidative enzymes in carcinoma of the large intestine in man. Am J Pathol 35:113–137.

Webb WA, McDaniel L, Jones L (1985) Experience with 1000 colonoscopic polypectomies. Ann Surg 201:626–632.

Wegener M, Borsch G, Schmidt G (1986) Colorectal adenomas: Distribution, incidence of malignant transformation, and rate of recurrence. Dis Colon Rectum 29:383–387.

Weiss H, Wildner GP, Jacobasch KH, Heinz U, Schaelicke W (1985) Characterization of human adenomatous polyps of the colorectal bowel by means of DNA distribution patterns. Oncology 42:33–41.

Welin S, Youker J, Spratt JS, Linnell F, Spjut HJ, Johnson RE, Ackerman LV (1963) The rates and patterns of growth of 375 tumors of the large intestine and rectum observed serially by double contrast enema study (Malmö technique). Am J Roentgenol 90:673–687.

West AB, Isaac CA, Carboni JM, Morrow JS, Mooseker MS, Barwick KW (1988) Localization of villin, a cytoskeletal protein specific to microvilli, in human ileum and colon and in colonic neoplasms. Gastroenterology 94:343–352.

Whitehead R, Skinner JM (1983) Tumor marker acquisition in the polyp to cancer sequence in the colon. Eur J Cancer Clin Oncol 19:923–927.

Wiebecke B, Brandts A, Eder M (1974) Epithelial proliferation and morphogenesis of hyperplastic adenomatous and villous polyps of the human colon. Virchows Arch A [Pathol Anat and Histol] 364:35–49.

Williams AR, Balasooriya BAW, Day DW (1982) Polyps and cancer of the large bowel: A necropsy study in Liverpool. Gut 23:835–842.

Williams CB, Macrae FA (1986) The St. Mark's neoplastic polyp follow-up study. Front Gastrointest Res 10:226–242.

Williams GT, Arthur JF, Bussey HJR, Morson BC (1980) Metaplastic polyps and polyposis of the colorectum. Histopathology 4:155–170.

Williams GT, Blackshaw AJ, Morson BC (1979) Squamous carcinoma of the colorectum and its genesis. J Pathol 129:139–147.

Willson J, Bittner G, Oberley T, Meisner L, Weese J (1987) Cell culture of human colon adenomas and carcinomas. Cancer Res 47:2704–2713.

Winawer SJ, Fleisher M, Green S, Bhargava D, Leidner SD, Boyle C, Sherlock P, Schwartz MK (1977) Carcinoembryonic antigen in colonic lavage. Gastroenterology 73:719–722.

Winawer SJ, Ritchie MT, Diaz BJ, Gottlieb LS, Stewart ET, Zauber A, Herbert E, Bond J (1986) The National Polyp Study: Aims and organization. Front Gastroint Res 10:216–225.

Winawer SJ, Zauber A, Diaz B, O'Brien M, Gottlieb LS, Sternberg SS, Waye JD, Shike M, National Polyp Study Work Group (1988) The National Polyp Study: overview of program and preliminary report of patient polyp characteristics. In G Steele, RW Burt, SJ Winawer, JP Karr (eds): Basic and Clinical Perspectives of Colorectal Polyps and Cancer. Alan R Liss, New York, pp 23–33.

Woda BA, Forde K, Lane N (1977) A unicryptal colonic adenoma, the smallest colonic neoplasm yet observed in a nonpolyposis individual (letter to editor). Am J Clin Pathol 68:631–632.

Wolley RC, Schreiber K, Koss LG, Karas M, Sherman A (1982) DNA distribution in human colon carcinomas and its relationship to clinical behavior. J Natl Cancer Inst 69:15–22.

Yonezawa S, Nakamura T, Tanaka S, Maruta K, Nishi M, Sato E (1983) Binding of *Ulex europaeus* agglutinin-I in polyposis coli: Comparative study with solitary adenoma in the sigmoid colon and rectum. J Natl Cancer Inst 71:19–24.

Yonezawa S, Nakamura T, Tanaka S, Sato E (1982) Glycoconjugate with *Ulex europaeus* agglutinin-I-binding sites in normal mucosa, adenoma, and carcinoma of human large bowel. J Natl Cancer Inst 69:777–785.

Yuan M, Itzkowitz SH, Ferrell LD, Fukushi Y, Palekar A, Hakomori S, Kim Y (1986) Comparison of T-antigen expression in normal, premalignant, and malignant human colonic tissue using lectin and antibody immunohistochemistry. Cancer Res 46:4841–4847.

Yuan M, Itzkowitz SH, Ferrell LD, Fukushi Y, Palekar A, Hakomori S, Kim Y (1987) Expression of Lewis[x] and sialylated Lewis[x] antigens in human colorectal polyps. J Natl Cancer Inst 78:479–488.

Zotter St, Lossnitzer A, Hageman PhC, Delemarre JFM, Hilkens J, Hilgers J (1987) Immunohistochemical localization of the epithelial marker MAM-6 in invasive malignancies and highly dysplastic adenomas of the large intestine. Lab Invest 57:193–199.

3
Adenomatous Polyposes

These autosomal dominant conditions, including familial polyposis coli (FPC) and its variants, account for < 1 percent of the cases of colorectal carcinoma in the population (0.2 percent in Finland according to Mecklin, 1987). (See Figure 3.1.)

A. Familial Polyposis Coli and Gardner's Syndrome

Much of the basic data on FPC has been supplied by Bussey and colleagues (Bussey, 1975) at St. Mark's Hospital in London, where there is a Polyposis Register of over 300 families. Additional information, especially on the extracolonic manifestations, has been supplied by workers in Europe, the United States, and Japan (see Haggitt & Reid, 1986).

The original definition of FPC was an autosomal dominant disorder characterized by multiple adenomatous polyps of the colon. The minimal number of polyps was established at 100, since the number of polyps counted by Bussey in 39 cases ranged from 104 to > 5,000, the average being about 1,000. None of the non-familial ("multiple polyps") cases approaches this number, and it is now believed that subjects with between 50 and 100 APs probably also have FPC. The disease occurs in 1/7,000 to 1/24,000 live births. Penetrance has been variously estimated at 80–100 percent. About one third of the cases have no family history and are assumed to represent new mutants, who can, parenthetically, transmit the disease to their offspring. This high mutation rate may be an overestimate, resulting from incomplete history taking, late onset of disease, or incomplete gene expression/penetrance.

In 18 colectomy specimens from patients with FPC, polyp density was left > right > transverse (Bussey, 1975). The vast majority of polyps are tubular adenomas < 0.5 cm with very rare hyperplastic polyps. There is no evidence that the pathological features of APs in FPC differ from those of sporadic APs. The mean age of detection of APs in symptomatic individuals is 36 years and in screened asymptomatic subjects is 24 years. The disease probably starts in the first decade of life when small (1–2 mm) mucosal nodules can be seen

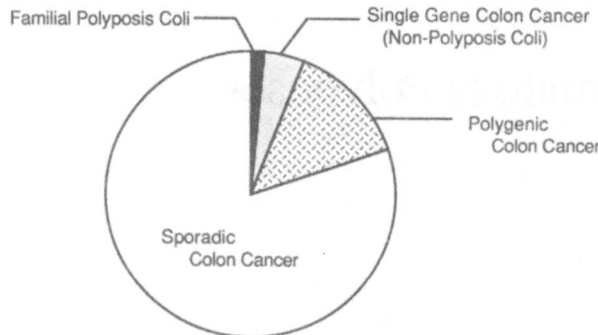

FIGURE 3.1. Diagrammatic presentation of the contribution of nonpolyposis colon cancer to the colon cancer burden. (Slightly modified from Lynch et al., 1985.)

endoscopically and when biopsies of even flat mucosa may show early adenomatous change (see Lev et al., 1987). The mean age of diagnosis of carcinomas in FPC is 39 years. The incidence of carcinoma increases progressively with time and is virtually 100 percent by age 55. Multiple carcinomas are very common, being synchronous in 41 percent and metachronous in 7 percent. Adenomas in terminal ileum have been described following colectomy for FPC/Gardner's syndrome (GS) (Hamilton et al., 1979).

In the early 1950s, Gardner and his colleagues described a variety of extraintestinal disorders associated with adenomatous polyps of the colon (Gardner & Richards, 1953). These included epidermoid cysts, fibromas, and osteomas; in the following decades other lesions were added to GS, including various fibromatoses, dental abnormalities, gastric and small intestinal polyps and carcinomas (primarily periampullary carcinoma), and lymphoid hyperplasia of terminal ileum. The majority of the gastric polyps were found to be hamartomatous, and the duodenal polyps were mainly adenomas (Yao et al., 1977; Sarre et al., 1987). Tumors of the hepatobiliary and endocrine systems have also been described.

A variety of abnormalities have been described in cultured cutaneous fibroblasts and epithelial cells in Gardner's syndrome and FPC. A 98 percent increase in tetraploidy was found in GS patients and a 15 percent increase in patients with FPC (Danes, 1981); the fibroblasts in both conditions were also much more susceptible to chemical or viral-induced transformation than were normal cutaneous fibroblasts. Other abnormalities of fibroblasts include a deformed cytoskeleton, elevated plasminogen-dependent protease production, and a loss of anchorage dependence. Some of these characteristics have suggested to Kopelovich (1980) the possibility that such fibroblasts are "initiated" cells in accord with the initiation-promotion theory of carcinogenesis (Berenblum, 1941). He suggested further that the disorder of fibroblasts may be systemic and affect the cells of the pericolonic fibroblast sheath and thus be responsible for the abnormal growth of colonic epithelium.

Certain biochemical disturbances are also seen in morphologically normal colonic mucosa in subjects with FPC/GS. These include retention by surface epithelium of ability to synthesize DNA (Deschner & Lipkin, 1975) and increased levels of ornithine decarboxylase (Luk & Baylin, 1984). The level of this enzyme, which is involved in polyamine synthesis, is low in colonic mucosa from normal individuals but is increased in proliferating lesions such as APs. The flat mucosa from FPC subjects, like APs, shows staining for *Ulex europaeus*, a fucose-binding lectin that has been used as a marker for blood group substance H(O), whereas normal mucosa from nonpolyposis individuals fails to so stain (Yonezawa et al., 1983).

There is some evidence that Gardner's syndrome and FPC represent variants of the same disorder. The pathology of the colonic polyps is the same in both. Of 200 families with FPC, Bussey found that 18 percent had two to nine members with some features of GS. In a radiological survey, 90 percent of patients with FPC had radiopaque mandibular lesions characteristic of Gardner's syndrome in the absence of other features of GS (Utsunomiya & Nakamura, 1975), further blurring the distinction between the two disorders. Others such as McKusick (1974), however, feel that they represent two separate entities. In the experience of Naylor and Lebenthal (1980), the two syndromes tend to breed true: The offspring of Gardner's syndrome patients will eventually manifest extracolonic pathology, whereas those of FPC subjects will not. The marked difference in tetraploidy in cutaneous fibroblasts from the two conditions (cited above) also supports the two-entity theory.

B. Other Adenomatous Polyposis Syndromes

Turcot's syndrome (Turcot et al., 1959) represents a variant of FPC exhibiting central nervous system (CNS) tumors, mainly glioblastomas. Unlike FPC, it has an autosomal recessive mode of inheritance (Erbe, 1976). Rare cases of Torre (or Muir) syndrome, a familial disorder characterized by sebaceous gland tumors and colorectal and other visceral carcinomas, show colonic polyposis (see Alessi et al., 1985). Some workers, including Lynch et al. (1981) feel that the Torre syndrome is a subtype of the cancer family syndrome (see Chap. 4.F.4), however, which does not exhibit polyposis.

Enterline (1976) has suggested the term *adenomatous polyposis* followed by description of any additional lesions in an attempt to curb the proliferation of eponyms.

C. The Genetics of Adenoma Transmission

> And fate shall have no reprieve
> Euripides,
> *Hecuba*

FPC and Gardner's syndrome are autosomal dominant disorders. The mode of transmission of the far more common solitary or multiple (< 50) APs, presumed by most investigators to be precursors of the majority of the remaining cancers, is unclear. Veale (1965) believed that they have an autosomal recessive pattern resulting from a genotype he designated $pp-p$ representing the gene for sporadic adenomas; P, its allelic gene for FPC, and x, the normal gene for that locus. According to his theory, a high percentage of individuals in populations with high prevalence rates of APs will be heterozygous (px). Bussey et al. (1978) extended this hypothesis to account for the occurrence of colorectal cancers in the cancer family syndrome. They argued that the polyp-bearing children from homozygous (pp) unions will develop a high incidence of carcinomas, giving the false impression of a dominant inheritance of the carcinomas, whereas in reality the antecedent polyps were actually inherited according to a recessive mode. They speculated further that the occurrence in ulcerative colitis of dysplasia, a premalignant epithelium, and of adenomatous changes in juvenile polyps could be explained by the presence of the pp trait in patients with those diseases. However, local processes such as pathological repair in ulcerative colitis would appear to be more important than genetic considerations in explaining the genesis of dysplasia in that condition.

Burt et al. (1985) examined a large number of asymptomatic members of kindreds displaying a high incidence of carcinomas without any recognized inheritance pattern, that is, FPC or hereditary nonpolyposis colon cancer (HNCC). Their flexible sigmoidoscopic examinations revealed a significantly greater number and size of APs in the family members than in the spouse controls. Using a pedigree analysis that evaluated the likelihood of dominant, recessive, or random models, they found that the dominant mode best fitted their observations. Although this disagrees with the hypothesis of Veale that sporadic polyps follow a recessive mode, it confirms Veale's belief that it is the *predisposition toward* polyp formation that is inherited rather than the resulting carcinomas. They also suggested that an AP could represent an early phenotypic marker for subjects with genetically determined CRCs and that this could be investigated by applying chromosome mapping techniques, including restriction-fragment-length polymorphisms (RFLP), to that population. Further work from that group, drawing on the extensive Utah Genealogical Database (Cannon-Albright et al., 1988), examined larger numbers of families exhibiting colorectal cancer clustering and extended their earlier investigations to kindreds of subjects with sporadic APs discovered in (1) spouse controls during the earlier study and (2) the pathology files of their hospitals. Rectosigmoid adenomas were found in 19 percent of 407 relatives of the probands from all the "neoplastic" groups but only in 12 percent of the 263 new spouse controls ($p < 0.02$). The view that sporadic polyps may be genetically determined has also been expressed by McConnell (1980). A dissenting opinion was expressed by Grossman and Milos (1988), who performed colonoscopy in 154 subjects who had one or two first-degree relatives with CRC. They found an 18 percent prevalence of APs in that group, a rate even lower than they had calculated (24 percent) for the age-adjusted general population.

Some observers feel that the gene for Gardner's syndrome is different from that of FPC, one of the reasons for believing that the two disorders are distinct. In

Gardner's syndrome, it is uncertain as to whether there is a single pleiotropic gene with multiple manifestations or several closely linked genes: The former theory is favored by most.

Karyotypes of early adenomas in FPC (and in sporadic polyps) are normal. More dysplastic APs show abnormalities in chromosome numbers 8 and 14, preceding invasiveness (see Naylor & Lebenthal, 1980). These changes suggest that these APs harbor genetic instability similar to that exhibited by the cutaneous fibroblasts in FPC and GS (Danes, 1981; Gardner et al., 1982) and support the hypothesis that these adenomatous lesions are prone to increased somatic mutation and malignant transformation. Perhaps related to these findings is the deficient DNA repair noted in peripheral leukocytes from subjects with APs but not in those with hyperplastic polyps or normal colons (Pero et al., 1985).

Recently, two groups have shown *ras* gene mutations in villous adenomas and in 40 percent of colon carcinomas using hybridization with a variety of protooncogene probes (Bos et al., 1987; Forrester et al., 1987). They concluded that such mutations may provide the cells harboring them with a selective growth advantage. In many tumors, the same *ras* mutation was found in the adenomatous and carcinomatous components, suggesting that the mutation preceded the malignancy. Oncogene abnormalities such as amplification and deletion were also found by another group using similar methodology (Meltzer et al., 1987) in a villous adenoma and in 22 percent of colon carcinomas. They postulated that the increased cell proliferation in APs predisposes them to such oncogene abnormalities. Monnat et al. (1987) found that overexpression of certain oncogenes (*myc, fos, ras*) in CRCs was associated with more aggressive behavior. They also found that oncogene expression in polyps paralleled that of the synchronous carcinomas.

In the past few years, a series of investigations based on observations by Herrera et al. in 1986 have demonstrated partial deletion of the long arm of chromosome 5 in FPC, Gardner's syndrome, and 20 percent (or more) of sporadic CRCs (Bodmer et al., 1987; Solomon et al., 1987). Fearon et al. (1987) found loss of chromosome 17 sequences in 75 percent of 20 CRCs and in a few of the 30 APs studied using RFLP. They felt their data supported a monoclonal origin of colorectal tumors. Further collaborative efforts by several of the above groups (Vogelstein et al., 1988) showed (1) *ras* gene mutations in many adenomas > 1 cm in size and in carcinomas and (2) deletions in certain gene sequences in chromosomes 5, 17, and 18 in substantial numbers of adenomas and/or carcinomas. These authors developed a hypothesis of colorectal carcinogenesis based on their work and that of others. They suggested that the familial adenomatous polyposis (FPC) locus on chromosome 5 normally inhibits colonic epithelial proliferation and that its inactivation induces hyperproliferation in FPC and probably in patients with sporadic adenomas. A second event would be required to convert this epithelium to an adenoma. This could involve *ras* gene mutations or deletions in tumor suppressor genes in chromosomes 17 or 18 or other chromosomes, accompanied by DNA hypomethylation of their genomes, which is known to result in loss (or gain) of chromosomes. The central role of chromosome 5 in the genesis of sporadic carcinomas was questioned by Law et al. (1988). They found allelic loss on that chromosome in only 19 percent of 31 CRCs and in 2 percent of 42 adenomas, whereas allelic losses on chromosomes 17 and 18 were found in 56 percent and 52 percent of the CRCs, respectively. The significance of the various chromosomal changes and the role they may play in malignant transformation of APs have also been discussed in a recent review by Sandberg (1988).

References

Alessi E, Brambilla L, Luporini G, Mosca L, Bevilacqua G (1985) Multiple sebaceous tumors and carcinomas of the colon: Torre syndrome. Cancer 55:2566–2574.

Berenblum I (1941) The mechanism of cocarcinogenesis: A study of significance of cocarcinogenic action and related phenomena. Cancer Res 1:807–814.

Bodmer WF, Bailey CJ, Bodmer J, Bussey HJR, Ellis A, Gorman P, Lucibello FC, Murday VA, Rider SH, Scrambler P, Sheer D, Solomon E, Spurr NK (1987) Localization of the gene for familial adenomatous polyposis on chromosome 5. Nature 328:614–616.

Bos JL, Fearon ER, Hamilton SR, Verlaan-deVries M, van Boom JH, van der Eb AJ, Vogelstein B (1987) Prevalence of ras gene mutations in human colorectal cancers. Nature 327:293–297.

Burt RW, Bishop DT, Cannon LA, Dowdle MA, Lee RG, Skolnick MH (1985) Dominant inheritance of adenomatous colonic polyps and colorectal cancer. N Engl J Med 312:1540–1544.

Bussey H (1975) Familial Polyposis Coli. Johns Hopkins University Press, Baltimore.

Bussey HJR, Veale AMO, Morson BC (1978) Genetics of gastrointestinal polyposis. Gastroenterology 74:1325–1330.

Cannon-Albright LA, Skolnick MH, Bishop DT, Lee RG, Burt RW (1988) Common inheritance of susceptibility to colonic adenomatous polyps and associated colorectal cancers. N Engl J Med 319:533–537.

Danes BS (1981) Occurrence of in vitro tetraploidy in the heritable colon cancer syndromes. Cancer 148:1596–1601.

Deschner EE, Lipkin M (1975) Proliferative patterns in colonic mucosa in familial polyposis. Cancer 35:413–418.

Enterline HT (1976) Polyps and cancer of the large bowel. In B Morson (ed): Current Topics in Pathology: Vol 63. Pathology of the Gastrointestinal Tract. Springer-Verlag, Berlin, pp 95–141.

Erbe RW (1976) Inherited gastrointestinal-polyposis syndromes. N Engl J Med 294: 1101–1104.

Fearon ER, Hamilton SR, Vogelstein B (1987) Clonal analysis of human colorectal tumors. Science 238:193–197.

Forrester K, Almoguera C, Han K, Grizzle WE, Perucho M (1987) Detection of high incidence of K-ras oncogenes during human colon tumorigenesis. Nature 327:298–303.

Gardner EJ, Richards RC (1953) Multiple cutaneous and subcutaneous lesions occurring simultaneously with hereditary polyposis and osteomatosis. Am J Hum Genet 5: 139–147.

Gardner EJ, Woodward SR, Burt RW, Neff LK (1982) In vitro characterization of Gardner's syndrome, familial polyposis coli, hereditary discrete colorectal polyps and carcinoma. In RA Malt, RCN Williamson (eds): Falk Symposium 31: Colonic Carcinogenesis. MTP Press Ltd, Lancaster, pp 13–22.

Grossman S, Milos ML (1988) Colonoscopic screening of persons with suspected risk factors for colon cancer. Gastroenterology 94:395–400.

Haggitt RC, Reid BJ (1986) Hereditary gastrointestinal polyposis syndromes. Am J Surg Pathol 10:871–887.

Hamilton SR, Bussey HJR, Mendelsohn G, Diamond MP, Pavlides G, Hutcheon D, Harbison M, Shermeta D, Morson BC, Yardley JH (1979) Ileal adenomas after colectomy in nine patients with adenomatous polyposis coli/Gardner's syndrome. Gastroenterology 77:1252–1257.

Herrera L, Kakati S, Gibas L, Pietrzak E, Sandberg AA (1986) Gardner's Syndrome in a man with interstitial deletion of 5 q. Am J Med Genet 25:473–476.

Kopelovich L (1980) Hereditary adenomatosis of the colon and rectum: Recent studies on the nature of cancer promotion and cancer prognosis in vitro. In SJ Winawer, D Schottenfeld, P Sherlock (eds): Progress in Cancer Research and Therapy, Vol. 13. Raven Press, New York, pp 97–108. (Colorectal Cancer: Prevention, Epidemiology and Screening)

Law DJ, Olschwang S, Monpezat JP, Lefrancois D, Jagelman D, Petrelli NJ, Thomas G, Feinberg AP (1988) Concerted nonsyntenic allelic loss in human colorectal carcinoma. Science 241:961–965.

Lev R, Lebenthal E, Rossi T, Lance P (1987) Histochemical and morphological analysis of colonic epithelium from children with Gardner's syndrome and adults bearing adenomatous polyps. J Pediatr Gastroenterol Nutr 6:414–425.

Luk GD, Baylin SB (1984) Ornithine decarboxylase as a biologic marker in familial colonic polyposis. N Engl J Med 311:80–83.

Lynch HT, Lynch PM, Pester J (1981) The Cancer Family Syndrome. Rare cutaneous phenotypic linkage of Torre's syndrome. Arch Intern Med 141:607–611.

Lynch HT, Kimberling WJ, Albano W, Lynch JF, Biscone K, Schuelke GS, Sandberg AA, Lipkin M, Deschner EE, Mikol YB, Elston RC, Bailey-Wilson JE, Danes BS (1985) Hereditary nonpolyposis colorectal cancer (Lynch syndromes I and II) I. Clinical description of resource. Cancer 56:934–938.

McConnell RB (1980) Genetics of familial polyposis. In SJ Winawer, D Schottenfeld, P Sherlock (eds): Progress in Cancer Research and Therapy, Vol. 13. Raven Press, New York, pp 69–71. (Colorectal Cancer: Prevention, Epidemiology and Screening)

McKusick VA (1974) Genetics and large bowel cancer. Am J Dig Dis 19:954–958.

Mecklin J-P (1987) Frequency of hereditary colorectal carcinoma. Gastroenterology 93:1021–1025.

Meltzer SJ, Ahnen DJ, Battifora H, Yokota J, Cline MJ (1987) Protooncogene abnormalities in colonic cancers and adenomatous polyps. Gastroenterology 92:1174–1180.

Monnat M, Tardy S, Saraga P, Diggelman H, Costa J (1987) Protooncogenes and colon cancer. Int J Cancer 40:293–299.

Naylor EW, Lebenthal E (1980) Gardner's syndrome. Recent developments in research and management. Dig Dis Sciences 25:945–959.

Pero RW, Ritchie M, Winawer SJ, Markowitz MM, Miller DG (1985) Unscheduled DNA synthesis in mononuclear leukocytes from patients with colorectal polyps. Cancer Res 45:3388–3391.

Sandberg AA (1988) Chromosome studies in polyposis coli. In G Steele, RW Burt, SJ Winawer, JP Karr (eds): Basic and Clinical Perspectives of Colorectal Polyps and Cancer. Alan R Liss, New York, pp 203–213.

Sarre RG, Frost AG, Jagelman DG, Petras RE, Sivak MV, McGannon E (1987) Gastric and duodenal polyps in familial adenomatous polyposis: A prospective study of the nature and prevalence of upper gastrointestinal polyps. Gut 28:306–314.

Solomon E, Vass R, Hall V, Bodmer WF, Jass JR, Jeffreys AJ, Lucibello FC, Patel I, Rider
    SH (1987) Chromosome 5 deletions and colon cancer. Nature 328:616–619.
Turcot J, Despres JP, St. Pierre F (1959) Malignant tumors of central nervous system
    associated with familial polyposis and the colon: Report of two cases. Dis Colon Rec-
    tum 2:465–468.
Utsunomiya J, Nakamura T (1975) The occult osteomatous changes in the mandible in
    patients with familial polyposis coli. Br J Surg 62:45–51.
Veale AMO (1965) Intestinal Polyposis. Eugenics Laboratory Memoirs, Series 40. Cam-
    bridge University Press, London.
Vogelstein B, Fearon ER, Hamilton SR, Kern SE, Preisinger AC, Leppert M, Nakamura
    Y, White R, Smits AMM, Bos JL (1988) Genetic alterations during colorectal-tumor
    development. N Engl J Med 319:525–532.
Yao T, Iida M, Ohsato K, Watanabe H, Omae T (1977) Duodenal lesions in familial poly-
    posis of the colon. Gastroenterology 73:1086–1092.
Yonezawa S, Nakamura T, Tanaka S, Maruta K, Nishi M, Sato E (1983) Binding of *Ulex
    europaeus* agglutinin-I in polyposis coli: Comparative study with solitary adenoma in
    the sigmoid colon and rectum. J Natl Cancer Inst 71:19–24.

4
Malignant Potential of Adenomatous Polyps

A. Association of Polyps and Carcinoma

Adenomatous polyps (APs) that contain invasive carcinoma will be discussed in Chap. 5.B, and the parallel occurrence of polyps and colorectal carcinomas (CRCs) in various population groups will be explored in Chap. 4.D. The coexistence of polyps and separate synchronous or metachronous (subsequent) carcinoma in the same individual will be discussed here. Although the epidemiological studies described in Chap. 4.D suggest a relationship between adenomatous polyps and carcinoma, they do not directly address the questions of (1) whether persons with adenomatous polyps are at elevated risk of subsequent CRC and, if so, (2) the size of the risk and its relation to pathological features of the polyp.

The premalignant nature of APs would be best demonstrated by the in vivo transition with time of a biopsy-proven benign polyp into a carcinoma. Since nonremoval of APs is clearly not allowable medically or ethically in the present era, this level of proof is rarely available. There are some studies of this type, dating mainly from the precolonoscopic period, when risk of surgical removal often exceeded the risk of malignant changes and the polyps were allowed to remain in situ and were followed radiographically. Many such studies were retrospective in design, frequently failed to define clearly the control population with which the study population was being compared, or had follow-up periods of insufficient length. In addition, p values or confidence intervals were often not included in the data analysis. Another problem is that interval endoscopic procedures were performed on some of these subjects; thus, the expected higher incidence of carcinoma might not be demonstrable in patients undergoing repeated polypectomies with removal of the precursor lesions (see also Gilbertsen & Nelms, 1978; Chu et al., 1986). If the original polyp were not biopsied, on the other hand, there may be uncertainty about its initial benignancy. Finally, the growth characteristics of "prevalent" lesions (polyps discovered by screening or at autopsy; carcinomas discovered by screening) may differ from the features of "incident" lesions (those discovered as a result of symptoms) (Morrison, 1985). Current large-scale prospective studies of polyp-bearing subjects being undertaken in Europe and the United States (Hermanek, 1985; Williams & Macrae, 1986;

TABLE 4.1. Factors increasing synchronous and metachronous cancer risk in patients with adenoma.

Villous histology
Severe dysplasia
Large size
Coexisting carcinoma
Adenoma adjacent to carcinoma
Multiple adenomas
Advanced age
Male sex

SOURCE: Slightly modified from Ekelund, 1980.

Winawer et al., 1986), some of which involve randomized trials of different surveillance schedules, will probably yield more reliable data.

1. Risk Factors for Carcinoma among Polyp-Bearers

Factors in patients and their polyps believed to be related to cancer risk are listed in Table 4.1. A future list of risk factors may also include a family history of colorectal neoplasia: most (but not all) studies indicate that relatives of subjects with APs or CRCs show increased rates of both lesions (e.g., Cannon-Albright et al., 1988; see Burt and Samowitz, 1988; Grossman et al., 1989).

Dysplasia in a polyp found incidentally at colectomy for a first primary carcinoma is felt to be a marker for metachronous cancer (Morson et al., 1983). A recent prospective pilot study, however, found no association between dysplasia in a polyp (without synchronous carcinoma) and metachronous carcinoma (Williams & Macrae, 1986). Long-term prospective studies with larger numbers of patients may help resolve this important question. Less controversial factors include size and number of original polyps, which have been shown to correlate well with the risk of subsequent cancer (Figure 4.1). Earlier work had shown a relationship between number of adenomatous polyps and cancer risk (see Morson et al., 1983). Those authors found that the risk rose from 22 percent in subjects bearing one polyp to 69 percent for those with six or more polyps. In FPC, where hundreds or thousands of polyps are present, the risk approaches 100 percent.

2. Natural History of Polyps; Metachronous Carcinomas

Reports of malignant transformation of small numbers of adenomatous polyps were reported in the earlier literature. In the four cases quoted by Potet and Soullard (1971), transformation took 7, 8, 8, and 15 years, but the original diagnosis of benignancy was made on radiological rather than on histological grounds. Muto et al. (1975) found malignant transformation in 3 of 4 adenomatous polyps after 5, 6, and 13 years, although only 1 of the 3 had an initial (benign) biopsy.

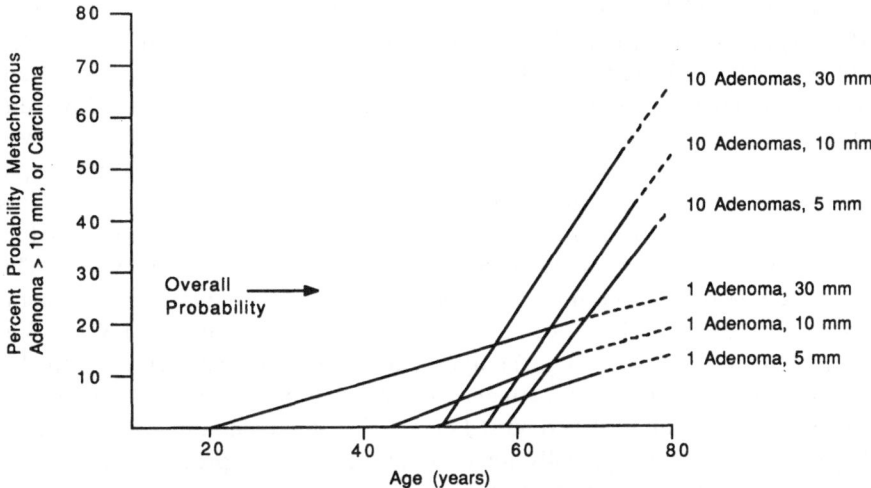

FIGURE 4.1. Relationship between patient age, the number and size of adenomas at poly-pectomy, and the probability of finding a large adenoma or carcinoma at follow-up. (Slightly modified from Williams & Macrae, 1986 with permission of S. Karger AG, Basel.)

They also noted transformation in 2 of 10 villous adenomas (VAs) after 10 and 28 years, but it is unclear if there was initial biopsy proof of benignancy in the 2 transformed VAs. On the other hand, in one study of 300 polyps ranging from 0.2 to 4.0 cm in size followed for two to nine years (Figiel et al., 1965), none became malignant, at least on radiological appearance, during that period, although several showed rapid growth necessitating removal. The follow-up period in that study was relatively short, however. Cohorts of patients with FPC who had no treatment were followed for > 20 years by Muto et al. (1975). At 0–5 years, 12 percent had developed carcinoma, but between 15 and 20 years, 56 percent had. Of course these figures cannot be extrapolated to patients with sporadic polyps.

More recent studies either have included a control group followed in parallel to the polyp-bearers or have compared the results in the study group with the CRC incidence in the age- and sex-adjusted general population, often from the same geographical area. Brahme et al. (1974) followed 115 patients with adenomatous polyps and 113 polyp-free individuals for 10.5 and 9.8 years, respectively, excluding members of the polyp group who developed CRC within 5 years of the initial diagnosis. They found three cases of CRC in the polyp group and none in the controls. Prager et al. (1974) assembled 305 patients with polyps at the Lahey Clinic and followed 283 of them for at least 15 years or to death. Carcinoma was found in 12 patients, whereas the expected number was 6.5 for a risk of about 2. A case control study has also suggested an association between CRC and a history of adenomatous polyps (Jain et al. 1980). In that study the risk for carcinoma in the polyp-bearers was estimated as 9.8 percent for males and 6.4 percent for females. Spencer et al. (1984) studied 751 patients with small (< 1 cm) polyps for an average of 13.5 years and found only a small excess of CRCs (18 versus 15.3 expected). Morson and Bussey (1985) did a retrospective study of 1,001 polyp-bearing patients followed for 1–15 years after initial polypectomy. They found a cumulative risk for CRC of 2.5 percent after 5 years and of 6.7 percent after 15 years but provided no figures for a control population. The risk was higher for males, for subjects with multiple polyps, and after 15 years, for people whose polyps were > 1 cm. Lotfi et al. (1986) followed a total of 323 patients who had a polyp > 1 cm in diameter or a polyp < 1 cm that was biopsied or excised, for up to 25 years (48 percent were followed for at least 15 years). The observed rate of metachronous carcinoma was 2.7 times the rate in the general population (20 cases observed; 7.3 cases expected). The relative risk was 3.3 in the patients with excised (larger) polyps and 2.5 in the subjects with the smaller polyps. The risk was 5.3 in patients with multiple polyps and 1.4 for solitary polyps. Stryker et al. (1987) reported on a series of 226 patients with polyps > 1 cm followed radiographically for up to 20 years in the precolonoscopic era. They found that only 37 percent of the polyps showed detectable growth during the follow-up period. The cumulative incidence of carcinoma anywhere in the colon was 14 percent after 10 years and 35 percent after 20 years; the latter included a 24 percent incidence at the site of the original polyp and 11 percent elsewhere. Although no comparison figures were given, this incidence is clearly higher than in the general population.

The most extensive follow-up study is that of Gilbertsen and Nelms (1978), who examined > 18,000 subjects over a 25-year period following the initial polypectomy. Since repeated polypectomies were performed during that period, this cannot be considered a study of the natural history of adenomatous polyps. The results, however, have been used as support for the AP–carcinoma sequence since only 15 percent of the expected number of rectosigmoid carcinomas were found. A full critique of this important investigation will be given in Chap. 5.A.

To summarize, the weight of evidence indicates that persons with adenomatous polyps of the colon, or at least with polyps > 1 cm, are at elevated risk of future CRC (as well as of recurrent adenomatous polyps). The study of Brahme et al. (1974) is too small to provide a meaningful estimate of the magnitude of the carcinoma risk. The larger studies by Prager et al. (1974) and Lofti et al. (1986) suggested a relative risk of about 2–3. The study by Spencer et al. (1984) did not show much of a carcinoma risk, but that study was intentionally restricted to small polyps. Little information on CRC *mortality* risk or on risk in relation to histopathological features of polyps is provided by these studies.

3. Peak Incidences of Adenomatous Polyps and Colorectal Carcinomas

Another more indirect way to estimate the length of the putative AP–cancer sequence is to compare age of peak incidences for the two lesions. Muto et al. (1975) found this difference to be 4 years for the sporadic cases and 12 years for FPC. Grinnell and Lane (1958) found figures in the literature of 8–10 years for sporadic cases. Although the age of peak incidence of cancer is fairly close to

TABLE 4.2. Age adjusted incidence of cancers and polyps of the colon and rectum in a standard population of 1 million.

Age groups	Number of cancers	Cumulative % of cancers	Number of polyps	Cumulative % of polyps	Ratio of cancer to polyps
20–40	21	4	11,300	18	1/540
40–49	50	15	9,000	33	1/180
50–59	98	38	15,000	58	1/150
60–69	124	65	15,000	83	1/120
70–79	120	91	8,500	97	1/70
80 +	40	100	1,500	100	1/40
Totals	453		60,300		1/130

SOURCE: Modified from Spratt et al., 1958.

the actual appearance of that tumor, the polyps may have been present for many years prior to their detection, especially in symptomatic patients with large polyps. Some of these intervals may thus be gross underestimates. Kozuka et al. (1975) demonstrated a progressive increase in average age of patients with increasing grades of dysplasia in APs from 44 years (the least dysplasia) to 56 years (invasive carcinoma), with intermediate values for patients with moderate and severe dysplasia. Large standard deviations and absence of p values mar that report, however.

Rough estimates of the AP–cancer interval based on the data described above and in A.2 indicate that in those individuals whose polyps are destined to become malignant, the process takes an average of 10 years. Preliminary results from the National Polyp Study suggest that in high-risk subjects the interval from normal mucosa to adenoma is five years and from adenoma to CRC is another five years (Winawer et al., 1988). All observers agree, on the other hand, that most adenomatous polyps never become malignant. The AP/carcinoma ratio remains elevated even in older age groups where carcinomas are relatively more common (Table 4.2).

4. Metachronous Carcinomas in Subjects with Synchronous Polyps and Carcinomas

Muto et al. (1975) found that the presence of a synchronous AP in a subject with carcinoma, which occurs in one third or more of cases, raises the cumulative risk of a second carcinoma in the following 25 years to 10 percent; the risk is 4 percent when no polyp is present. Similar results have been found by other groups such as Chu et al. (1986), who have reviewed the literature and documented the increase in both synchronous and metachronous carcinomas in their own series of subjects with carcinomas with and without synchronous polyps (Table 4.3). They felt that the number and size of the initial polyps correlated better with synchronous than with metachronous carcinomas, since removal of those polyps removed a possible source of metachronous carcinomas.

TABLE 4.3. Incidence of synchronous (SC) and metachronous (MC) carcinoma (%) in 1,202 patients with colorectal carcinoma.

	Synchronous polyps	
	No	Yes
SC	0.7	11.0*
MC, 5 yr	1.6	3.9*
MC, 10 yr	3.4	6.5*

*Compared with no synchronous polyps: $p < 0.001$.
SOURCE: Modified slightly from Chu et al., 1986.

5. Prevalence of Polyps in Subjects with Colorectal Carcinomas

The prevalence of polyps, especially when multiple, is higher among subjects bearing carcinomas than those who do not. This has been demonstrated by numerous groups (e.g., Grinnell & Lane, 1958; Bockus et al., 1961; Muto et al., 1975; Eide, 1986). In one surgical study of 591 CRCs, adenomas more commonly accompanied right-sided than left-sided carcinomas (47 percent versus 22 percent, $p < 0.001$), and synchronous APs and carcinomas occurred more frequently in older and in male subjects (Slater et al., 1988). Adenomas in patients with carcinomas tended to be larger and to display villous features and severe dysplasia more frequently than did adenomas in carcinoma-free subjects. Also, clustering of adenomas within the same or adjacent colonic segments, clustering among multiple carcinomas, and clustering of adenomas and carcinomas have been observed (Blatt, 1961; Eide & Schweder, 1984) and cited as further evidence for the relationship between the two lesions.

6. Opposing Views

Other arguments have been raised by those who do not believe in the adenoma–carcinoma sequence, in addition to the lack of overlap between polyp and carcinoma distribution (the reasons for which will be discussed in Section D.2). One such argument is the nonrandom distribution of solitary polyps around solitary carcinomas in subjects bearing both lesions (Spratt et al., 1958). Those authors reasoned that if one of two polyps became malignant, this should occur with equal frequency distally and proximally to the polyp that remains benign. Since this proved not to be the case, they concluded that this constituted evidence against the malignant potential of polyps. I do not find this argument convincing. All their results show is that some additional factor may account for why the more proximal of the polyps in the right colon and the more distal of the two polyps in the left colon tend to become malignant and not that adenomatous polyps do not become malignant at all. Another argument used by the same authors against the polyp–carcinoma sequence is the lack of statistically significant difference between polyp prevalence in carcinoma-containing and carcinoma-free colons in their study. The relative risk for polyps in subjects bearing carcinomas between the ages of 50 and 70 was found by them to be about 1.4, which is indeed quite low. The risk found by others, however, is higher, ranging from 2.0 (Bockus et al., 1961) to 4.7 (Eide, 1986). Although the latter figure is probably too high, as conceded by the author, a rough estimate from the cumulative literature indicates that the average risk of polyp occurrence in carcinoma-bearing patients is about two to three times normal and even higher for multiple polyps, as indicated above. This must thus be considered a definite, if low, risk and supportive of an association between polyps and carcinomas.

B. Histological Evidence of Malignant Transformation

In many adenomatous polyps, transitions between, or juxtaposition of, glands showing varying degrees of dysplasia and between these and invasive carcinoma can be readily found. This has been amply documented in the literature (Helwig, 1947; Grinnell & Lane, 1958; Lescher et al., 1967; Potet & Soullard, 1971; Kozuka et al., 1975; Muto et al., 1975; Shinya & Wolff, 1979) and is illustrated in Figures 2.11, and 2.12 and in the frontispiece. Others agree that this may occur with villous or tubulovillous adenomas but not with the much more common tubular adenoma (Spratt et al., 1958; Castleman & Krickstein, 1962; Spjut & Estrada, 1977; Pozharisski & Chepick, 1978). Some of these authors, for example Castleman and Krickstein, feel that definite stalk invasion is very rare in true tubular adenomas, especially pedunculated ones, that the carcinoma in situ described by others in such adenomas represents atypia only, and that tubular adenomas showing even these changes are "dead-end" lesions that only very rarely, if ever, metastasize or exhibit other biological characteristics of malignancy. Disagreement as to what constitutes villosity and invasiveness may partly explain these differences of opinion. A study designed to quantify the frequency of carcinomas arising in tubular adenomas was carried out by Kaneko (1972). Of 458 adenomatous polyps (VAs were excluded), carcinoma was found in 4 percent of TAs and in 20 percent of TVAs; his results are somewhat clouded by the inclusion of 5 "noninvasive" carcinomas in his total of 43 malignant polyps, however. The results are nevertheless similar to those from the larger subsequent series of Muto et al. (1975) in which 4.8 percent of 1,880 purely TAs and 22.5 percent of 383 TVAs showed invasive carcinoma. My personal view is that even purely tubular adenomas may show clear-cut stalk invasion and that this is not such an exceptional event (see fig. 1 in Lev, 1979).

C. Adenomatous Remnants in Carcinomas

Man sieht, was Man weiss.
Goethe,
Schriften zur Kunst

Another way to investigate the adenoma–carcinoma relationship is to examine invasive carcinomas for residual adenoma. This has been found in 14–23 percent of carcinomas (Muto et al., 1975; Eide, 1983). Adenomatous remnants are more frequently present in carcinomas only infiltrating the submucosa (57 percent) than in those extending beyond the muscularis propria (8 percent). The implication is that as carcinomas grow and infiltrate, they destroy preexisting adenomatous tissue. Remnants are also more frequent when the carcinoma is well differentiated and when it is exophytic. Carcinomas found in patients with additional synchronous APs or with multiple carcinomas are more likely to contain remnants than are carcinomas occurring as solitary lesions. Some investigators, however, find no remnants of tubular adenomas in carcinomas (Spratt et al., 1958; Castleman & Krickstein, 1962). The inability to detect tubular remnants could result from identification of such remnants as villous or as well-differentiated carcinoma, or from failure to take blocks for the "most benign" area of the carcinomas.

Bussey (1978) found adenomatous remnants more frequently in carcinomas complicating FPC (36 percent) than in sporadic carcinomas from the same institution (11 percent). He ascribed this to the earlier stages of some of the synchronous carcinomas in FPC, it being well known (see above) that remnants are more readily identified in carcinomas showing superficial infiltration only.

D. Epidemiological Evidence and Etiological Factors

The general prevalence, geographic variation, and anatomic location of APs have been described in Chapter 2.

1. Sex and Age

All studies indicate higher prevalence of APs in males, the typical male-to-female ratio being around 3:2 in several large series (Rider et al., 1954; Grinnell & Lane, 1958; Konishi & Morson, 1982). The sex ratio is higher than that found in CRC (Correa et al., 1977; Clark et al., 1985). The prevalence increases with age (Clark et al., 1985), as does that of CRC. In Western societies, a significant number of polyps is found over age 50, the usual age targeted for screening for colorectal carcinoma. Thus, a polyp prevalence of 41 percent was found in American men between 50 and 54 years in the autopsy study of Rickert et al. (1979) and of at least 50 percent in subjects over 60 years in other studies of similar American populations (Chapman, 1963; Figiel et al., 1965).

2. Relationship of Polyps and Colorectal Carcinomas

The prevalence of adenomas at autopsy has been found to be positively associated with the incidence rate of CRC in various populations (Correa et al., 1977; Restrepo et al., 1981; Segal et al., 1981; Bat et al., 1986; see also Table 2.1). This applies to populations in different countries, to different groups within the same country (e.g., Ashkenazi and non-Ashkenazi Jews in Israel; whites and blacks from the same area of South Africa), and to members of the same racial group who have migrated from a country with a low incidence of both diseases to a high-incidence country (Japanese in Japan versus Japanese in Hawaii or in the United States). Risk factors for carcinoma such as large size, significant villous component, and severe dysplasia are also more common in polyps in countries with high polyp prevalence.

Exceptions do exist. There were fewer polyps at autopsy in Liverpool than in Norway despite the higher incidence of cancer in Liverpool (Eide & Stalsberg, 1978; Vatn & Stalsberg, 1982; A.R. Williams et al., 1982). Equal numbers of polyps were found in two separate areas of Norway having markedly different cancer incidence. In these instances, populations with lower-than-expected adenoma prevalence had larger polyps than populations with lower cancer incidence. This indicates that in some groups risk factors such as large size bear a closer relationship than polyp prevalence to cancer incidence.

It has been found in autopsy studies that polyps are evenly distributed throughout the colon, whereas carcinomas tend to be concentrated in the left colon. This lack of anatomic overlap between the two lesions has been used as an argument against the malignant potential of APs. If one examines not the polyp prevalence but the presence of risk factors for malignancy in polyps, most studies indicate that such factors are more common in left-sided than in right-sided polyps (Grinnell & Lane, 1958; Arminski & McLean, 1964; Ekelund & Lindstrom, 1974; Gillespie et al., 1979; Shinya & Wolff, 1979; Granqvist, 1981; A.R. Williams et al., 1982; Isbister, 1986; Matek et al., 1986; Richards et al., 1987; Winawer et al., 1988). Others have found no difference in degree of dysplasia (Stemmermann & Yatani, 1973; Eide & Stalsberg, 1978), size (Rickert et al., 1979; Vatn & Stalsberg, 1982), or percentage of villosity (Christie, 1981) between left- and right-sided polyps. The lack of congruence between these risk factors and carcinoma distribution noted by these latter authors may have several sources. In the Rickert study, subjects with previous resections for carcinomas were excluded, which may have eliminated the larger synchronous and metachronous polyps in the rectosigmoid area from the counts. Although Christie found a lack of predominance of villous lesions in the left colon, malignant polyps were more common there, suggesting that factors other than villosity may be prime determinants for malignant changes in APs in some populations. The relatively high prevalence of cecal carcinomas noted by some authors (e.g., Spratt et al., 1958) may also correlate to some extent with the relatively high frequency of right-sided villous polyps noted by Christie. The lack of predominance of dysplasia among left-sided polyps noted by the two groups cited above is more difficult to explain. It is possible that in those two populations dysplastic right-sided polyps progress more slowly or less frequently into carcinoma than do left-sided ones, or that other risk factors such as large size play a more important role in malignant transformation in such distal polyps.

3. Dietary and Other Factors

Diets high in animal fat and/or low in fiber are suspected of predisposing people to CRC (Ron & Lubin, 1986). These and other dietary factors have been invesigated in relation to APs of the colon. Hoff et al. (1986) compared dietary intakes of 78 persons with polyps and 77 persons without polyps. People with polyps > 5 mm had a somewhat lower intake than polyp-free persons of fiber, carbohydrate, cruciferous vegetables, vitamin C, and iron and higher intakes of total fat and animal protein. Macquart-Moulin et al. (1987) obtained dietary histories from 252 subjects with and 238 subjects without polyps. Polyps were inversely related to carbohydrate and fiber (the carbohydrate effect was stronger than the fiber effect) and showed a positive association with sugar intake and saturated fat (weak effect). In a prospective study of dietary and other factors in Hawaii Japanese (Stemmermann et al., 1988), 163 individuals came to autopsy including 79 with APs and 84 without. In contrast to the above studies, no significant differences were found between the two groups in intake of dietary fat, protein, or carbohydrate.

If a diet high in saturated fat is a cause of colonic polyps, this condition might be correlated with atherosclerosis. A positive correlation was found in the autopsy studies of Correa et al. (1982) in which men but not women showed the association and of Stemmermann et al. (1986) who investigated Hawaii-Japanese men.

A diet high in saturated fat increases the fecal excretion of bile acids and their metabolites, abetted by the increased activity of certain anaerobic bacteria such as *Clostridium*. It is believed, largely from data obtained from animal experiments, that elevated fecal bile acids act as colon tumor promoters (see Reddy & Wynder, 1977); the increased proliferative activity of colonic epithelium may result from increased ornithine decarboxylase activity or activation of protein kinase C. Those authors compared the excretion of bile acids and neutral sterols between patients with APs and controls. Levels of deoxycholic and lithocholic acids and of cholesterol, coprostanol, coprostanone, and cholestane-3β, 5δ, 6β-triol were higher in polyp-bearers, but cholic and chenodeoxycholic acids, β-sitosterol, and campesterol were similar in both groups. Other workers have not found fecal neutral steroids or bile acids to be increased in subjects with APs, however (Tanida et al., 1984; Peuchant et al., 1987). Van der Werf et al. (1982) found colonic absorption of deoxycholate to be elevated in patients with polyps.

Because of the increased fecal concentration of certain cholesterol metabolites in individuals with polyps, serum cholesterol has also been analyzed in that population. Neugut et al. (1986) found the mean serum cholesterol to be nearly identical in 248 people with APs and in 688 controls, whereas Mannes et al. (1986) found a small positive association of cholesterol levels and polyps in their study of 155 patients with and 687 patients without polyps. The estimated prevalence of APs in the latter study was nearly twice as high in persons with cholesterol > 269 mg/dl than in persons with cholesterol < 177 mg/dl.

Some investigators have found increased numbers of right colon carcinomas in women with a history of remote *cholecystectomy*, possibly the result of alterations in fecal bile acids after that procedure. It is unclear if cholecystectomized individuals also develop increased numbers of APs, some groups having found an increase, whereas others have not. An intriguing explanation for a lack of increase in APs in the face of increased CRC has been offered by Neugut et al. (1988). They suggested that cholecystectomy might be a risk factor not for the development of new polyps but rather for the progression of APs into CRCs (see factors E_2 and C in section D.4 below).

A positive association of APs and *alcohol* consumption was noted in a sigmoidoscopic study (Diamond, 1972) and in a recent prospective autopsy study (Stemmermann et al., 1988). In a screening study, Hoff et al. (1987) found that APs were increased in cigarette *smokers* and suggested that smoking might have an initiating effect on colorectal mucosa; Stemmermann et al. (1988), however, found no relationship between smoking history and polyps.

A brief review of dietary and nondietary factors was recently carried out by Stemmermann (1989). Of all the variables examined, which included factors such as body weight, serum cholesterol, and protein and fat intake, the only one showing a statistically significant correlation with AP prevalence was alcohol intake.

4. Etiology

The genetics of adenoma transmission in FPC and in sporadic APs has already been discussed in Chapter 3; the new information on *ras* gene mutations and chromosome deletions (Chap. 3.C) has been invoked to explain the various stages of carcinogenesis. The interplay between genetic and environmental effects has been well summarized by Morson et al. (1983): "The prevalence of adenomas in a population is determined by environmental factors but genetic factors determine which members of a uniformly exposed population actually develop the lesion" (p. 472).

The relationships between colonic adenomas and carcinoma may be viewed as shown in Figure 4.2. If adenomas are premalignant, their causes (C_1) will also affect the rate of carcinoma (Figure 4.2a). Such agents may be somewhat easier to identify by the study of adenomas than by the study of carcinoma (Morrison, 1979; Hoff et al., 1987). On the other hand, a given cause of carcinoma (C_2) is not necessarily associated with the rate of adenomas (Figure 4.2b). Furthermore, the two conditions could share a cause (C_3) even if adenomas are not premalignant (Figure 4.2c). Obviously, more complex relationships can be envisioned.

Another approach to etiological factors has been suggested by Hill (1978) that is based on the assumption that the majority of CRCs originate in APs. He postulated that three environmental factors (E_1, E_2, and C) were responsible for (1) the appearance of adenomas (E_1) in genetically susceptible subjects (see Veale, 1965 and Chap. 3.C), (2) their growth into larger adenomas (E_2), and (3) the transformation of the latter into carcinoma (C). Factor C would also account for the small number of CRCs originating in nonadenomatous cells. He felt that it was necessary to invoke three factors because of the marked variation in the proportions of large APs in different countries and the low incidence of CRC in some countries such as Finland that have a high-fat diet normally associated with a high carcinoma incidence. He suggested that the low incidence in Finland could be due to

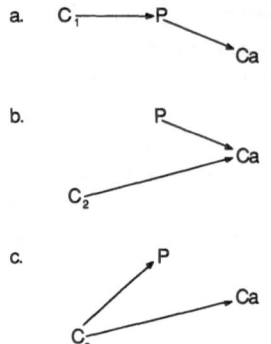

FIGURE 4.2. Some possible interrelations of causes and progression of neoplasia of the colon. C_1 = cause of polyps; C_2 = cause of cancer; C_3 = cause of polyps and cancer; P = polyps; Ca = cancer. (Courtesy of A.S. Morrison)

a low level of E_1 or low numbers of genetically prone subjects. More recent studies have indicated that the high-fiber intake in Finland may neutralize the high dietary fat content (thereby transforming a high E_1 factor into a low E_1 factor) and account for the low carcinoma incidence.

The concept of three separate environmental factors has been challenged by Berg (1988). He felt that a sole environmental factor was involved in the cause of adenomas but that once they appeared, whether in Americans, Japanese, or others, they had equal chances of becoming malignant. This hypothesis cannot account for the higher percentages of risk factors for carcinoma (e.g., large size) in polyps from populations with higher prevalence rates of APs, however.

E. Findings in Experimental Animals

> If we could pass sigmoidoscopes
> On tigers, wolves and antelopes;
> On cock-a-toos and hooded crows;
> On porcupines and buffaloes
> And other creatures—great and
>   small—
> We might discover that they all,
> In spite of contradictory rumors,
> Were prone to intestinal tumors!
> So much unlike outside yet much the
>   same within!
> Once more—"one touch of nature
>   makes the whole world kin!"
>   C.E. Dukes (1959)

Despite the above suggestion, most work on animals has involved chemically induced rather than spontaneously occurring tumors. The seminal work of Laqueur and colleagues must be acknowledged here. He found that cycad meal fed to rats (for other purposes) induced colon carcinomas in some animals (Laqueur et al., 1963) and that the carcinogenic component was a derivative of methylazoxymethanol, a compound still in popular use in experimental carcinogenesis.

1. The Observations

Several reviews have discussed the mode of action of chemical carcinogens, the effects of dietary and other environmental agents on carcinogenesis, and the relationship between chemically induced tumors in animals and spontaneous human tumors (see "Carcinogenesis and biochemical mechanisms," 1975; LaMont & O'Gorman, 1978; Goldin, 1988). Most of the work has been done in rodents by means of parenteral injection or intrarectal instillation of carcinogens such as dimethylhydrazine or N-methyl-N-nitrosourea. The type of tumor (adenoma or carcinoma) induced by these carcinogens varies according to species. Rats, for example, generally develop adenocarcinomas and a few APs, whereas mice show relatively more APs (Maskens & Dujardin-Loits, 1981). In addition, the same carcinogen may induce both tumors in the same species. In many of the models the distribution and histology of both tumors are similar to what is found in humans.

Many of the carcinogens also have toxic effects on the crypt cells, which are noted within several hours. The crypts recover from this damage within several days, but the toxic effects will reappear each time the drug is given. In many—but not all—models, mucosal hyperplasia is seen later in the course of drug administration prior to the appearance of tumors (Wiebecke et al., 1973; Chang, 1978; Maskens & Dujardin-Loits, 1981; Wargovich et al., 1983). This hyperplasia stage, which is not seen in humans, may represent a compensatory response to the mucosal damage described above. Although it temporally precedes the onset of neoplasia, it is not felt to be of histogenetic importance in carcinoma formation.

The preneoplastic stages have also been examined ultrastructurally (Fisher et al., 1981) and histochemically. There is a decrease in cytoplasmic mucus as cellular atypia appears and progresses. The residual mucin is of the sialomucin type (Filipe, 1975; Shamsuddin & Trump, 1981; Wargovich et al., 1983), which is believed by some (Filipe, 1975) to be a marker for premalignant change. Cellular DNA kinetics using tritiated thymidine shows reduced uptake during the early toxic period and increased uptake accompanied by increased labeling index and upward (surface) shift of the proliferative compartment during later periods (Wiebecke et al., 1973; Deschner et al., 1979; Wargovich et al., 1983). These latter changes resemble those found in human APs and in some normal mucosae from subjects with FPC or multiple APs.

The fully developed lesions consist of microscopic or gross foci of carcinoma or adenoma that may present as flat or polypoid tumors. As indicated above, the relative proportion of carcinomas to adenomas will depend on the species and strain of animal and to some extent on the histological criteria of the investigator.

There is a sharp difference of opinion as to whether transitions can be found between adenomatous and carcinomatous tissue. Some observers (Lev & Herp, 1978; Madara et al., 1983; Wargovich et al., 1983) readily find such transitions. In the study of Madara the percentage of carcinomas showing residual adenoma was found to be dependent on the degree of bowel wall infiltration, as is the case in humans (Muto et al., 1975). Others have not been able to demonstrate such transitions and feel that adenomas and carcinomas represent independent effects of carcinogens (Spjut & Spratt, 1965; Ward, 1974; Asano et al., 1975; Maskens & Dujardin-Loits, 1981; Shamsuddin & Trump, 1981). Some authors found evidence for both pathways, that is, carcinomas arising both de novo and from adenomas (Lev & Herp, 1978). In some animal models (e.g., Pozharisski, 1975) no adenomas at all were identified, the earliest microscopic lesions consisting of frankly carcinomatous glands.

2. Genesis of Tumors

Several groups have attempted to demonstrate histologically the earliest stages in tumor formation during chemical carcinogenesis. One investigator (Chang, 1985) concluded that there are two modes of proliferation of clones of dysplastic cells in the deep crypts. In one, there was upward extension of cells, which then accumulated at the crypt mouths owing to a presumed block of lateral migration by adjacent normal cells. In the other, there was budding of dysplastic cells into the mucosal lamina propria through defects in the epithelial basement membrane. These pathways are similar to those that are believed to occur during early adenoma formation in humans (Chap. 2.C; 2.G). Possible molecular events accompanying carcinogenesis have been suggested by Maskens (1981, 1983). He postulated two stages in the development of carcinomas in the rat based on a mathematical interpretation of cell transformation and tumor prevalence as a function of carcinogen dose and time. He felt that the first step represented a transmissible somatic mutation resulting from a carcinogen-induced change in DNA. The second event, which occurred in the presence or absence of the same carcinogen, could be epigenetic or mutational in nature. He speculated that the recessive somatic mutant of the first stage was converted to a homozygous state in the second stage.

3. Interpretation of Findings and Relevance to Humans

It is difficult to say how helpful the animal data are in understanding human colon carcinogenesis. The models have at least a superficial resemblance to human carcinomas in that left-sided tumors predominate and in that the histology of the adenomas and carcinomas is similar to their human counterparts. In contrast to the human situation, however, the experimental carcinomas predominate numerically over adenomas (at least in most of the rat models), the carcinomas tend to be multiple more often, and the large doses of carcinogens used to induce them bear little relation to any putative carcinogen in the human intestinal or external

TABLE 4.4. Risk of colorectal carcinoma in general population and in high-risk groups.

Group	Approximate lifetime cancer risk (%)
Normal population	5
Past history of breast or female genital cancer	7–20
Past history of colorectal cancer	15
Family history of colorectal cancer	15[a]
Adenomas	10–20[b]
Ulcerative colitis	5–50[c]
Cancer family syndrome (HNCC)	50
Familial polyposis coli	100

[a] Risk increases with number of relatives affected.
[b] Risk depends on number, size, and histology of adenomas.
[c] Risk depends on extent and duration of disease; the 50 percent figure applies to subjects with universal colitis of > 30 years duration.
SOURCE: Modified from Ron and Lubin, 1986, with permission of S. Karger AG, Basel.

environment. The animal models are quite useful, on the other hand, for testing the effects of diet, bacterial flora, injury, and other environmental factors on the frequency and timing of tumor induction.

Insofar as the histogenesis of ordinary human colorectal carcinoma is concerned, the usefulness of the various models described above may be limited. The tempo of carcinogenesis appears to be more rapid than in the human, even after compensating for the differences in life span between the two species, and it is possible that the early adenomatous phase in the animals is very brief and thus overlooked. It has been suggested that the rapidity of malignant change and the multiplicity of dysplastic foci are more reminiscent of human carcinomas arising in ulcerative colitis or FPC rather than in ordinary APs (Lev & Herp, 1978). Another problem in assessing the usefulness of the animal models is in the interpretation of the early mucosal changes, variously described as adenomatous, dysplastic, or malignant. Similar controversy clouds the definition of *residual* or *precursor tissue* in human malignant polyps (e.g., tubular adenoma versus well-differentiated adenocarcinoma) and interferes with a reproducible classification of the dysplastic epithelium in ulcerative colitis (Riddell et al., 1983; Collins et al., 1987). In fact, the same confusion and bias may be carried over from the human into the experimental field in that supporters of the de novo theory of human carcinogenesis tend to find evidence for that theory in their animal models, whereas those who believe that adenomas are precursors of human carcinomas find the same sequence in animals.

F. Alternate Hypotheses of Carcinogenesis

Individuals other than those with adenomatous polyps may be prone to develop CRC. The major high-risk groups are listed in Table 4.4.

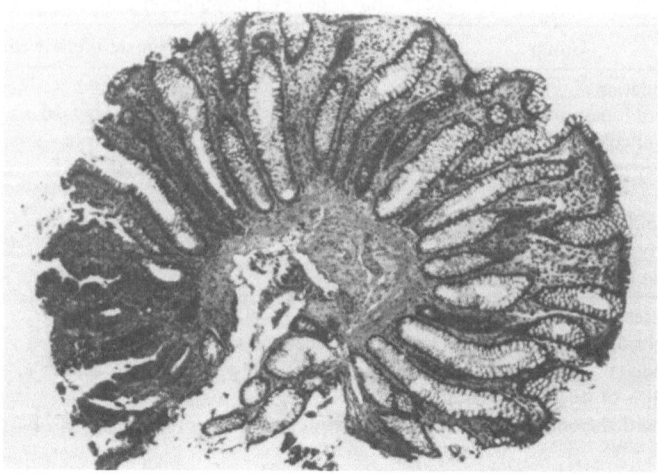

FIGURE 4.3. De novo carcinoma. Biopsy of abnormal, flat granular area in transverse colon from patient with previous adenomatous polyps and colorectal carcinomas. Dysplastic crypts are seen in lower left; the remaining crypts are slightly elongated but otherwise normal. H&E; ×35, reproduced at 100%.

1. The Flat Mucosa

The frequency with which cancers of the colon arise in the general population can be accounted for without assuming that any of them arose in adenomatous polyps. (Spratt & Ackerman, 1962, p. 132)

Despite unlimited opportunity to do so, minute or microcancer has not been observed in normal mucosa, i.e., unassociated with adenomatous tissue. (Lane, 1976, p. 2669)

Small numbers of early carcinomas have been described without apparent adenomatous components (Spratt & Ackerman, 1962; Kjeldsberg & Altshuler, 1970; Spjut et al., 1979; Crawford & Stromeyer, 1983; Kuramoto & Oohara, 1988). Some critics have accepted only those lesions < 1.0 cm or even < 0.5 cm as legitimate examples of de novo cancers, arguing that the larger and especially the ulcerated lesions may have destroyed preexisting adenomatous tissue (e.g., Lane, 1976). In fact, very few cancers in this smaller size range have been described in the literature. Even those, it has been claimed, may have evolved very rapidly through an adenomatous phase (see Lev & Herp, 1978), but this is a speculative objection that is difficult to prove. An example of a de novo carcinoma is shown in Figures 4.3 and 4.4. When removed 1.5 years later, following additional repeated biopsies, this flat indurated lesion 0.9 cm in diameter showed superficial submucosal invasion but no adenomatous remnants (Figure 4.5). Another grossly similar lesion from the same patient showed a few residual adenomatous glands when sectioned extensively, raising the possibility that some de novo carcinomas do in fact arise in adenomatous tissue (Figure 4.6).

Even less common are microscopic foci of carcinoma in grossly normal mucosa. These have been found in patients with FPC, sporadic CRCs, or no cancers

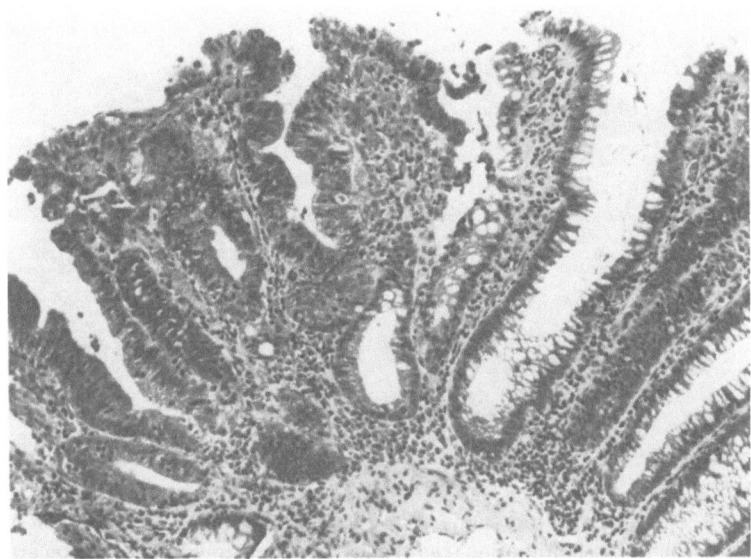

FIGURE 4.4. Higher magnification of severely dysplastic area shown in Figure 4.3. The crypts on the left show irregular contours with frequent buckling and are lined by basophilic cells showing loss of nuclear polarity and absence of mucus. No milder degrees of dysplasia suggestive of preexisting adenomatous epithelium were found (compare with Figure 4.6). Crypts on right show mildly reactive changes with enlarged nuclei and reduced numbers of goblet cells. H&E; ×176, reproduced at 55%.

(Bussey, 1975; Pozharisski & Chepick, 1978; Lev & Grover, 1981; Shamsuddin, 1982). Such de novo lesions are very rare when compared with the frequency of microscopic adenomatous foci in the flat mucosae in the same populations (Bussey, 1975; Lev & Grover, 1981; see also Rickert et al., 1979; Muto et al., 1980). In addition to their rarity, questions can be raised as to the malignant nature of some of these foci (e.g., fig. 1 in Shamsuddin, 1982). Even if all such minute ($\leq$ 0.2 cm) flat or polypoid lesions in the cumulative literature are accepted as de novo cancers, they do not begin to approach the frequency with which minute malignant foci in this size range are found in preexisting APs. Thus, few contemporary investigators are in accord with the conclusions of Spratt and Ackerman cited in the quotation heading this section or with Pozharisski and Chepick (1978) who claim to find consistently foci of carcinoma in situ in the flat mucosa and who believe that such lesions are the source of most invasive carcinomas.

2. Other Routes

a. Nonadenomatous Polyps

The most common of these is the hyperplastic polyp. Issues to be considered here are (1) whether such lesions are precursors of carcinomas and (2) whether they occur in colons prone to develop CRC.

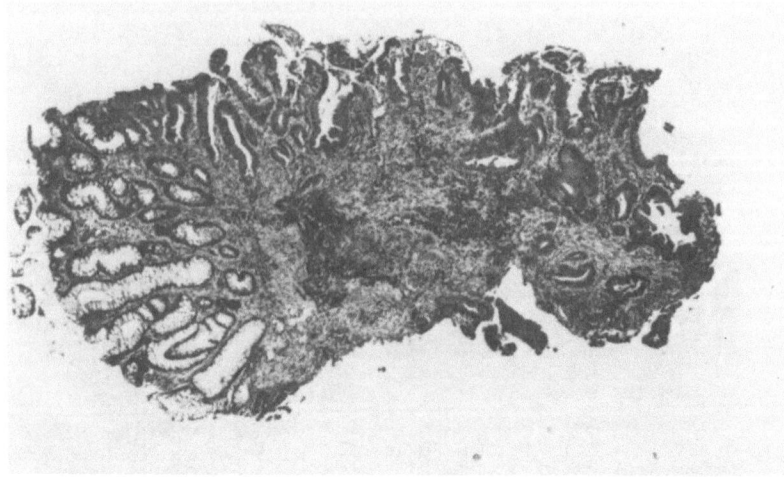

FIGURE 4.5. Early invasive carcinoma. Appeared 1.5 years later at site of lesion shown in Figure 4.3. Submucosal infiltration (right) and extensive stromal proliferation are noted. Multiple levels through this tumor failed to reveal adenomatous remnants. H&E; ×28, reproduced at 55%.

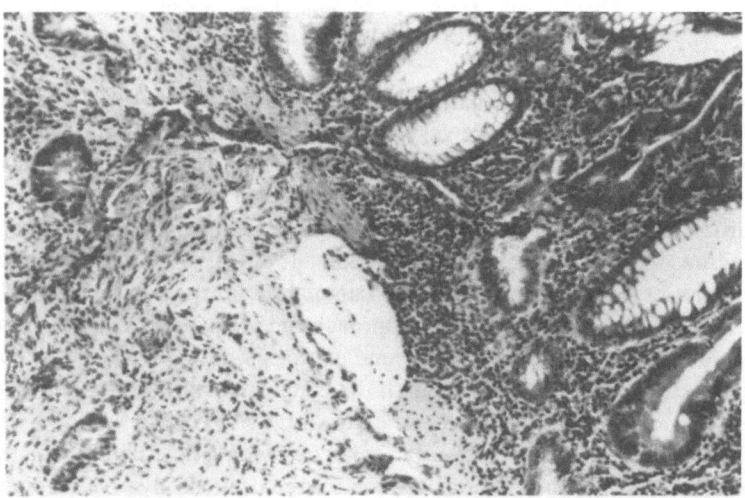

FIGURE 4.6. De novo carcinoma showing early submucosal invasion on left. Multiple sections revealed the mildly dysplastic adenomatous gland in the mucosa in the lower right. Glands like these are believed to represent the origin of the severely dysplastic glands in the upper right and, by inference, of the invasive carcinoma. It is possible that serial sections of other de novo carcinomas would reveal more such adenomatous remnants. H&E; ×176, reproduced at 100%.

There is very little evidence to indicate that HyPs evolve into carcinomas. In contrast to APs, which frequently show foci of severe dysplasia and occasionally show invasive carcinoma, transitions between HyPs and carcinoma are exceptionally rare, if they exist at all. In several large series, such transitions were not found (Spjut & Estrada, 1977; G.T. Williams et al., 1980). There are scattered reports of carcinoma arising in HyPs especially when they are multiple or > 1.0 cm (Cooper et al., 1979; Franzin et al., 1984; Bengoechea et al., 1987). Most of these HyPs also had adenomatous foci, and it is probable that the carcinomas arose in the adenomatous rather than in the hyperplastic areas, but at least one case of carcinoma arising in a strictly HyP has been described (Franzin & Novelli, 1982). It is clear that only a tiny fraction of CRCs arise in HyPs.

There are data indicating that some characteristics are shared by HyPs and neoplastic lesions, that is, APs and carcinomas, including increased expression of peanut agglutinin and carcinoembryonic antigen and decreased quantities of O-acylated sialomucins (Boland et al., 1982; Cooper & Reuter, 1983; Jass, 1983; Altavilla et al., 1984; Rhodes et al., 1986). Occasional staining of HyPs for oncofetal antigens, including modified blood group antigens, has also been described (Skinner & Whitehead, 1981; Bara et al., 1983; Cooper et al., 1987; Yuan et al., 1987), but the extent of reactivity is much less than in APs. Some workers feel that the similarities with neoplasia are more striking in the larger HyPs (Skinner & Whitehead, 1981; Jass et al., 1984), but this has been disputed by others (Cooper et al., 1987). There are other data indicating that HyPs do *not* have properties associated with neoplasia. For example, one recent flow cytometry study showed that HyPs were always diploid, in contrast to APs, some of which showed aneuploidy and abnormalities of cell cycle fractions (Hoff et al., 1985).

Other evidence of anatomic and epidemiological nature points to an association between HyPs and carcinoma. HyPs are concentrated in the rectosigmoid colon, sites of high carcinoma prevalence, and are frequently found in the vicinity of such carcinomas (see Figure 2.17). The prevalence of HyPs is high in populations exhibiting high rates of CRC, and it rises in persons migrating from low CRC risk areas to high CRC risk areas.

The *conclusion* of these observations is that HyPs are found in populations at high risk for CRC and may in that sense serve as markers for carcinoma in such groups (see Jass, 1983). The HyP itself, however, only exceptionally, if ever, becomes malignant. The analogy with endometrial polyps comes to mind here: Although uteri harboring such polyps are at increased risk of developing carcinoma, the polyps themselves rarely become malignant. Subjects bearing APs, on the other hand, not only have an increased risk of having synchronous and developing metachronous carcinoma elsewhere in the colon but also have a definite chance of developing carcinoma in the polyp proper, depending on the presence of risk factors in that polyp as discussed above.

Other types of polyps occasionally become malignant. Such hamartomatous lesions as *Peutz-Jeghers polyps* and *juvenile polyps* were felt initially to have no malignant potential, with few exceptions (Horn et al., 1963), until the last decade. In this more recent period the tendency of patients with Peutz-Jeghers syndrome to develop carcinomas either at the site of the polyps, elsewhere in the colon, or in extracolonic sites has been repeatedly demonstrated (Linos et al., 1981; Burdick & Prior, 1982; Perzin & Bridge, 1982; Giardiello et al., 1987). Most of the gastrointestinal malignancies have been found in the stomach and upper small intestine. In some of these cases, adenomatous tissue was also identified in the polyps, or separate APs were found, and these may have been the source of the carcinomas. Similarly, patients with juvenile polyposis have been shown to be at slightly increased risk of developing CRC (Goodman et al., 1979; Lipper et al., 1981; Grotsky et al., 1982; Jarvinen & Franssila, 1984; Jones et al., 1987). Again, adenomatous tissue in the polyps and separate APs have been found in some of those patients, raising the possibility that the carcinomas may have originated in adenomatous lesions rather than in the hamartomas. It has been emphasized that carcinomatous change is more frequent when the polyps are multiple, and that polyps with multilobulated or villous features are more likely to have adenomatous (dysplastic) foci than are the more common spherical polyps (Jass et al., 1988). The contribution of these patients with carcinomas associated with hamartomatous polyps to the overall incidence of CRCs remains very small, however.

b. Inflammatory Bowel Disease

In *ulcerative colitis (UC)*, there is an increased risk of CRC, which develops in about 3–5 percent of all patients with that disease. The main risk factors are extent of colonic involvement, carcinoma occurring significantly more often in patients with universal colitics than in those with left-sided colitis, and duration of disease. Some workers feel that early onset of colitis is a separate risk factor, but others do not. The estimates of cumulative risk show variation due to geographic (Maratka et al., 1985) and methodological factors. The latter include type of population studied (e.g., private practice versus tertiary care), inclusion of patients with CRC at time of referral in some studies, follow-up periods that are incomplete or of inadequate length, and conclusions on long-term survivors based on very small numbers of patients (Sackett & Whelan ,1980; Collins et al., 1987). On the average, about 0.5 to 1 percent of patients with universal colitis develop carcinoma annually after 10 years of disease. Carcinomas arising in UC differ from those arising sporadically (Goldgraber & Kirsner, 1964; Cook et al., 1975; Greenstein et al., 1979). They occur in younger persons, are more often multiple and more evenly distributed throughout the colon, and often have unusual macroscopic appearances. Mucogenic and undifferentiated carcinomas are more frequent, although the latter are not quite as common as formerly believed.

In addition to frank carcinomas other gross abnormalities have been found in UC (Lennard-Jones et al., 1977; Yardley et al., 1979; Blackstone et al., 1981; Butt et al., 1983; Rosenstock et al., 1985). The dysplasia-associated lesion or mass was found by Blackstone et al. (1981) in 12 of 112 patients. They consisted of single or multiple polypoid lesions (10 cases), which were generally distinguishable endoscopically from pseudopolyps, and plaques (2 cases), of which 7 of 12 were associated with invasive carcinoma. More common, but of more controversial significance, are foci of microscopic dysplasia noted in random biopsies of flat mucosae. These were originally described in rectal biopsies by Morson and Pang (1967), who felt that patients bearing such lesions were at high risk for CRC. They suggested that a biopsy search for these dysplastic changes could identify a subset of colitics in whom prophylactic colectomy could be considered. In the 20-plus years since that study an enormous experience has accumulated based on that hypothesis; the more recent studies have been performed on colonoscopic biopsies. The carcinomas found at prophylactic colectomy were often of low Dukes stage, suggesting improved survival. Originally, mild, moderate, and severe grades of dysplasia were described, and the association with carcinoma was found to be closest with severe dysplasia. Other groups found that moderate dysplasia was not infrequently accompanied by carcinoma (Nugent et al., 1979; Lennard-Jones et al., 1983), prompting the suggestion that this degree of dysplasia also warranted colectomy.

In an attempt to provide a uniform and reproducible standardized nomenclature, a group of experts devised a new classification including only low- and high-grade dysplasia (Riddell et al., 1983) in which only the high-grade (severe) form should warrant consideration of colectomy. The subjective nature of several components of the grading system and the inclusion of an "indefinite for dysplasia" category may mitigate against its widespread adoption, however. Its usefulness has not yet been convincingly demonstrated in prospective studies of large numbers of cases. Another group of specialist pathologists, using a slightly different classification scheme, found only "fair" to "moderate" agreement over their four diagnostic categories; there was also a low level of agreement on the diagnosis of high-grade dysplasia between certain pairs of observers (Dixon et al., 1988).

The overall concept that identification of dysplasia can identify early and curable carcinomas has several flaws, some of which have been discussed in a recent review by Collins et al. (1987). The presence and degree of dysplasia may be difficult to determine. There is a 4–8 percent interobserver variation among experienced gastrointestinal pathologists on the grading of dysplasia, and this figure will surely be higher among the rank and file pathologists responsible for evaluating the majority of biopsies. The relationship between severe dysplasia and carcinoma is not as close as initial reports indicated. Such studies, which were almost always nonblinded, indicated that on the average > 90 percent of colitics with carcinoma also showed severe dysplasia in the colectomy specimen. A carefully designed study in which slides were examined in a blinded fashion indicated that only 50 percent of such colectomies showed severe dysplasia distant from the carcinoma (Ransohoff et al., 1985). It has also been shown that only 45 percent of colitics with severe dysplasia on biopsy show carcinoma in the subsequent colectomy specimen. It is thus clear that there are deficiencies in both sensitivity and specificity of the severe dysplasia factor. Finally, although early reports indicated that the majority of carcinomas discovered in prophylactic colectomy specimens performed for dysplasia were Dukes A, cumulative experience shows that < 30 percent of carcinomas are in this category. Collins et al. (1987) concluded that colonoscopic surveillance for dysplasia in UC is only rarely beneficial and is not cost-effective.

Morphometric techniques based primarily on nuclear characteristics are said to help differentiate between low- and high-grade dysplasia in UC (Allen et al., 1987), although that study revealed wide ranges of values for the low-grade groups with some overlap with cases of carcinoma. Those authors emphasized the importance of proper field selection and of integrating the morphometric and histological findings. They subsequently examined architectural features (such as area of epithelium, lamina propria, and mucosa) and the number of crypts per unit length of muscularis mucosae (Allen et al., 1988) and were able to separate normal and regenerative from dysplastic epithelium but still observed slight overlap between low- and high-grade dysplasia. DNA analysis has also been performed on UC specimens. In a flow cytometric study, Hammarberg et al. (1984) found aneuploidy in 25 percent of dysplastic specimens and in 10 percent of inflamed mucosae. In a cytophotometric (in situ) study, Cuvelier et al. (1987) were able to differentiate between low-grade dysplasia, characterized by slight hyperploidy, and high-grade dysplasia, in which hypertetraploidy and aneuploidy were frequently found.

Markers for dysplasia in UC have been investigated in an attempt to refine the indications for colectomy. Increased amounts of sialomucins and increased expression of peanut agglutinin, a CRC-associated lectin, have been found in goblet and columnar cells in dysplastic and nondysplastic mucosae from patients with UC (Boland et al., 1984; Ehsanullah et al., 1985; Pihl et al., 1985; Cooper et al., 1987; Fozard et al., 1987). The usefulness of both of these markers has been disputed by Jass et al. (1986), who found no differences in their staining intensities in colitics with and without carcinoma. Another group investigated staining for carcinoembryonic antigen and the secretory immunoglobulin system in nondysplastic, dysplastic, and malignant colonic epithelium from patients with UC, using an immunoperoxidase technique (Allen et al., 1985). They found that the variable expression of those antigens in the above mucosae precluded their use in differentiating benign from dysplastic epithelium. Ahnen et al. (1987) also found no qualitative differences between staining intensities or distribution of several mucins and lectins or of CEA in dysplastic and nondysplastic foci. Tumor-associated antigens have been found in mucus extracts and tissue sections from both involved and noninvolved mucosae in UC (Haviland et al., 1988). However, these antigens were found in only some of those subjects, and as conceded by the authors, epitope expression in involved mucosae could have resulted from nonspecific inflammation and tissue damage. It thus appears that the above tissue markers, while worthy of continued study and of potential interest in understanding neopolastic transformation in UC, have little current use in the clinical management of patients with UC. A scanning electron microscopic study showed reduction in the number of surface cells and in the width of microvilli in dysplastic epithelium as compared with nondysplastic epithelium, and the authors suggested that this technique might supplement light microscopy in the diagnosis of dysplasia (Shields et al., 1985).

Little is known of the *histogenesis* of CRC in UC. In contrast to sporadic carcinomas, which are believed to arise frequently in APs, carcinomas in UC are believed to arise in the flat mucosa. Such tumors presumably result from what has been termed *pathological repair*, that is, a chronic regenerative response to repeated bouts of mucosal damage. It is possible that the increased numbers of dividing epithelial cells are more susceptible to carcinogens than normal colonic cells. The exact site of malignant transformation is not known, although kinetic studies have revealed expansion of the proliferative compartment, as is also the case in FPC.

CRC is also known to complicate *Crohn's disease*, the incidence of carcinoma being 4- to 20-fold that found in the control population (Weedon et al.,1973; Gyde et al., 1980; Hamilton, 1985). The absolute rate of carcinoma is about threefold less than in UC (Greenstein et al., 1980). The carcinomas have many of the same epidemiological and pathological characteristics as those found in UC (Hamilton, 1985). In addition, severe dysplasia has been noted in mucosal biopsies from patients with Crohn's colitis (Craft et al., 1981; Simpson et al., 1981; Petras et al., 1987). Colonoscopic surveillance for dysplasia in Crohn's colitis is of more questionable benefit than in UC because of the lower incidence of carcinoma, the as-yet-unknown incidence of dysplasia, and the presence of carcinoma in excluded loops and in the small intestine in Crohn's disease (Butt & Morson, 1981; Simpson et al., 1981).

c. Miscellaneous

APs and carcinomas of the sigmoid colon are known to occur at sites of ureterosigmoidostomy anastomoses performed for urinary diversion on patients with carcinoma or various benign conditions of the urinary bladder. The interval between surgery and the subsequent colon carcinoma is on the average about 20 years. The incidence of carcinomas is between 100- and 550-fold more than that of the general population (Labow et al., 1979; O'Higgins et al., 1981). Possible etiological factors include postresection hyperplasia of the sigmoid mucosa, the irritating effects of urine, or the presence in urine of a carcinogen or an enzyme activating a fecal procarcinogen (Bristol & Williamson, 1981). Although this procedure was replaced by ileal conduit in the early 1950s, it is still occasionally performed, and sigmoid tumors should be anticipated in such patients.

In certain parts of China, CRCs secondary to *Schistosoma japonicum* have been found and the genesis of CRC compared with that following UC (Ming-Chai et al., 1965). CRCs associated with schistosomiasis or amebiasis have also been described in Africa (Adekunkle & Abioye, 1980). CRCs have been found in areas of radiation-induced damage, generally many years after substantial exposure (> 3,000 rads) to radiotherapy administered usually for gynecological malignancies (see Rotmensch et al., 1986).

It has been claimed that the transitional mucosa adjacent to CRCs shows specific premalignant changes consisting of increased sialomucins and decreased sulfomucins (Filipe & Branfoot, 1976). If proven to be correct, this theory would be of great interest since it suggests that an extended field of colonic mucosa is susceptible to malignant transformation. Several groups of investigators have found, however, that these histochemical changes occur in the vicinity of lesions other than primary colon carcinomas, such as metastases and diverticulitis, and they believe that such changes are nonspecific (Isaacson & Attwood, 1979; Listinsky & Riddell, 1981; Lev et al., 1985). Histological and ultrastructural abnormalities have been described in transitional mucosa, but these too are probably nonspecific. Dysplastic cells resembling those seen in well-established premalignant lesions such as APs and UC have not been found in this mucosa.

3. The Field Effect

In several organs, neoplasias or their precursors do not develop as solitary lesions but in multiple sites. It is not uncommon to find foci of dysplasia or carcinoma in situ in esophageal or uterine cervical mucosa distant from the frank carcinoma. In the colon it has been generally assumed that this is not the case, since only rarely do random sections of grossly normal mucosa in the vicinity of APs and carcinomas reveal microscopic neoplasia, and the theory (Filipe & Branfoot, 1976) that specific premalignant histochemical changes are found in this "transitional" mucosa (see 2.c above) has not been substantiated. Evidence that such mucosal areas may nevertheless be at increased risk for neoplasia is provided by the tendency of synchronous APs and carcinomas to cluster in the same colonic segment (Chap. 4.A.5). There are also biochemical and kinetic alternations in such mucosae, as discussed in Chaps. 2.F and 3.A. It has been recently shown that normal mucosae from widely separate colonic locations from patients with one or more polyps or with carcinoma show increased labeling index with tritiated thymidine, upward expansion of the proliferative compartment, and labeling of superficial epithelial cells that is rarely observed in normal mucosa (Terpstra et al., 1987; Ponz de Leon et al., 1988). The abnormalities were greater in mucosae from subjects with larger polyps or with carcinomas than in those with smaller polyps. No such changes were found in Crohn's colitis. A summary of the various phenotypic abnormalities in colonic mucosa from subjects at increased risk for CRC has been provided in a recent review by Lipkin (1986). Although these characteristics are of great biological interest, none has yet proved useful in the clinical management of this high-risk population.

4. Hereditary Nonpolyposis Colorectal Cancer

HNCC is discussed here since it is believed to account for up to 5 percent of CRCs. These kindreds show early onset of colorectal carcinoma (mean age 46 years), predominance of proximal colonic tumors (72 percent), multiple primary tumors, better survival than stage-matched sporadic CRCs, and autosomal dominant transmission (Lovett, 1976; Albano et al., 1982; Lynch et al., 1985; Mecklin, 1987; Lynch et al., 1988). Penetrance is 89 percent. HNCC comprises Lynch syndrome I (hereditary site-specific colon cancer), which has no extra-colonic tumors, and Lynch syndrome II (cancer family syndrome, or CFS), in which endometrial, breast, ovarian, and other carcinomas occur in addition to the colonic tumors. Synchronous CRCs are seen in 18 percent of subjects with HNCC and metachronous CRCs in 24 percent, both rates being significantly above those found in patients with sporadic CRCs.

It is presumed that the colonic carcinomas in this condition arise from the flat mucosa, since colonic polyposis is not part of the syndrome. However, APs, some of which show transitions with the carcinomas, have been found in affected patients (Lynch et al., 1983; Love & Morrissey, 1984). In fact, in one recent study, 64 percent of 17 patients with HNCC had APs (Swaroop et al., 1987), and in another study "flat" adenomas exhibiting lateral spread and concentrated in the right colon were described (Smyrk et al.,1988). Thus, the same adenoma–carcinoma sequence may be operative here as in the nonhereditary cases of CRC and in FPC.

G. Conclusions

1. Evidence of Malignant Potential of Adenomatous Polyps

Data on the association of APs and carcinomas in the same individual or in high-risk populations constitute only indirect evidence of malignant potential. Studies of the fate of polyps left in situ (natural history of polyps) have generally dealt with small numbers of patients and/or insufficient follow-up periods and have only rarely documented malignant transformation. The most convincing evidence has been (1) the histological demonstration of varying grades of dysplasia and of invasive carcinoma in the same polyp, and (2) the demonstration of adenomatous remnants in early carcinomas. This evidence has been accepted for VAs and TVAs and for the TAs in FPC but has been disputed in the case of the common sporadic TA. Disagreement centers on conflicting interpretations of epithelial atypia, villous structures, and invasiveness in such adenomas and on the frequency and nature of adenomatous remnants in small carcinomas. It is unlikely that further morphological studies will resolve this issue (see also discussions in Chap. 4.B and C above). Nevertheless, the majority opinion at the present time is that even tubular adenomas have the capacity to undergo malignant transformation, especially in the presence of such risk factors as multiplicity, large size, and severe dysplasia.

Both opponents and proponents of the adenoma–carcinoma sequence agree that the majority of APs never become malignant. Can this rare event account for most of the CRCs – or not? Most investigators feel that the majority of CRCs arise in APs, but there are few data on the relative contributions of APs, and of the de novo carcinomas and other nonadenomatous sources (e.g., other types of polyps), to the overall incidence of CRC in the population. In a recent attempt to construct a mathematical model to help determine screening schedules for subjects at increased risk of CRC, the assumption was made that 93 percent of CRCs originate in APs (Eddy et al., 1987), but it is unclear how those authors arrived at that figure. The American Cancer Society (1980) has estimated that 50–75 percent of CRCs develop from APs.

2. Significance of Malignant Potential of Adenomatous Polyps

Recognition of the adenoma–carcinoma sequence as a common event would have significant public health implications. It would suggest that resources should be allocated toward the identification and screening of populations for polyps and not only for carcinomas. The sensitivity and specificity of the various screening modalities are different for polyps and carcinomas (see Chap. 5.A); polyp detection might require a greater emphasis on endoscopy than on fecal occult blood testing, for example. Such a shift in screening policy would be predicated on the assumption that removal of precursor APs will reduce the incidence of CRC in such individuals, but this has not yet been convincingly demonstrated (see next chapter).

On the other hand, the chance that a detected AP is or will become malignant is no longer a factor in determining whether to perform polypectomy, at least for lesions > 5 mm. In the precolonoscopic era this risk had to be balanced against the morbidity and mortality of laparotomy, colotomy, and polypectomy, whereas currently polypectomy can be performed colonoscopically, a procedure associated with a very low morbidity and mortality.

3. The Problem of the Metachronous Carcinoma

The issue to be considered here is the probability of a future CRC developing in an individual who has just undergone complete polypectomy for one or more APs. In contrast to subjects in whom polyps were left in situ, in which cases future CRCs developed in those polyps and elsewhere (e.g., Stryker et al., 1987), individuals undergoing adequate polypectomies are only at risk for CRCs elsewhere in the colon. This risk will determine the surveillance schedule necessary for such persons. Information here is based largely on retrospective studies. Despite the defects of those studies, as discussed in this chapter, they do suggest that polyp-bearing subjects have a two- to threefold excess risk of developing metachronous carcinoma compared with normal subjects. More data on the magnitude of this risk may be provided by Polyp Registries and long-term prospective studies of subjects with APs. Of course, this will require the inclusion of a control population in whom the frequency of CRCs can be compared with that of polyp-bearers; use of spouse controls (Burt et al., 1985; Cannon-Albright et al., 1988) may deserve greater consideration in this regard. Assuming the occurrence of enough carcinomas, these studies should tell us not only how frequently such carcinomas occur in polypectomized individuals but also, hopefully, whether they arise in recurrent APs and how the characteristics of the initial polyps affect the risk for future carcinoma. One potential problem with many of these studies is that the interval polypectomies may interfere with the interpretation of some of the end results, as was the case with the retrospective studies: Since recurrent APs will be removed, there will be no way of knowing how many of them would have become malignant. It is possible that close colonoscopic surveillance and repeated polypectomies may actually reduce the incidence of metachronous carcinomas in these patients, but it will take many years to determine this. If a decrease in CRC incidence and especially CRC mortality are demonstrated, widespread adoption of such surveillance schedules may be indicated.

References

Adekunkle OO, Abioye AA (1980) Adenocarcinoma of the large bowel in Nigerians: A clinicopathologic study. Dis Colon Rectum 23:559–563.

Ahnen DJ, Warren GH, Greene LJ, Singleton JW, Brown WR (1987) Search for a specific marker of mucosal dysplasia in chronic ulcerative colitis. Gastroenterology 93: 1346–1355.

Albano WA, Recabaren JA, Lynch HT, Campbell AS, Maillard JA, Organ CH, Lynch JF, Kimberling WJ (1982) Natural history of hereditary cancer of the breast and colon. Cancer 50:350–363.

Allen DC, Biggart JD, Orchin JC, Foster H (1985) An immunoperoxidase study of epithelial marker antigens in ulcerative colitis with dysplasia and carcinoma. J Clin Pathol 38:18–29

Allen DC, Hamilton PW, Watt PCH, Biggart JD (1987) Morphometrical analysis in ulcerative colitis with dysplasia and carcinoma. Histopathology 11:913–926.

Allen DC, Hamilton PW, Watt PCH, Biggart JD (1988) Architectural morphometry in ulcerative colitis with dysplasia. Histopathology 12:611–621.

Altavilla G, Lanza G, Rossi S, Cavazzini L (1984) Morphologic changes, mucin secretion, carcinoembryonic antigen (CEA) and peanut lectin reactivity in colonic mucosa of patients at high risk for colorectal cancer. Tumori 70:539–548.

American Cancer Society (1980) Guidelines for the cancer-related checkup. Recommendations and rationale. CA 30:194–240.

Arminski TC, McLean DW (1964) Incidence and distribution of adenomatous polyps of the colon and rectum based on 1,000 autopsy examinations. Dis Colon Rectum 7: 249–261.

Asano T, Pollard M, Madsen DC (1975) Effects of cholestyramine on 1,2-dimethyl-hydrazine induced enteric carcinoma in germ-free rats. Proc Soc Exp Biol Med 150: 7880–7885.

Bara J, Languille O, Gendron MC, Daher N, Martin E, Burtin P (1983) Immuno-histochemical study of precancerous mucus modification in human distal colonic polyps. Cancer Res 43:3885–3891.

Bat L, Pines A, Ron E (1986) Colorectal adenomatous polyps and carcinoma in Ashkenazi and non-Ashkenazi Jews in Israel. Cancer 58:1167–1171.

Bengoechea O, Martinez-Penuela JM, Larrinaga B, Valerdi J, Borda F (1987) Hyperplastic polyposis of the colorectum and adenocarcinoma in a 24-year-old man. Am J Surg Pathol 11:323–327.

Berg, JW (1988) Epidemiology, pathology and the importance of adenomas. In G Steele, RW Burt, SJ Winawer, JP Karr (eds): Basic and Clinical Perspectives of Colorectal Polyps and Cancer. Alan R Liss, New York, pp 13–21.

Blackstone MO, Riddell RH, Rogers BHG, Levin B (1981) Dysplasia-associated lesion or mass (DALM) detected by colonoscopy in long-standing ulcerative colitis: An indication for colectomy. Gastroenterology 880:366–374.

Blatt LJ (1961) Polyps of the colon and rectum: Incidence and distribution. Dis Colon Rectum 4:277–282.

Bockus HL, Tachdjian V, Ferguson LK, Mouhran Y, Chamberlain C (1961) Adenomatous polyp of colon and rectum; its relation to carcinoma. Gastroenterology 41:225–232.

Boland CR, Lance P, Levin B, Riddell RH, Kim YS (1984) Abnormal goblet cell glycoconjugates in rectal biopsies associated with an increased risk of neoplasia in patients with ulcerative colitis: Early results of a prospective study. Gut 25:1364–1371.

Boland CR, Montgomery CK, Kim YS (1982) Alterations in human colonic mucin occurring with cellular differentiation and malignant transformation. Proc Natl Acad Sci USA 79:2051–2055.

Brahme F, Ekelund GR, Norden JG, Wenckert AM (1974) Metachronous colorectal polyps: Comparison of development of colorectal polyps and carcinomas in persons with and without histories of polyps. Dis Colon Rectum 17:166–171.

Bristol JB, Williamson RCN (1981) Ureterosigmoidostomy and colon carcinogenesis. Science 214:351.

Burdick D, Prior JT (1982) Peutz-Jeghers syndrome: A clinicopathologic study of a large family with a 26-year follow-up. Cancer 50:2139–2146.

Burt RW, Bishop DT, Cannon LA, Dowdle MA, Lee RG, Skolnick MH (1985) Dominant inheritance of adenomatous colonic polyps and colorectal cancer. N Engl J Med 312:1540–1544.

Burt RW, Samowitz WS (1988) The adenomatous polyp and the hereditary polyposis syndromes. In GD Luk (ed): Colorectal Cancer. Gastroenterology Clinics of North America. Volume 17, WB Saunders, Philadelphia, pp 657–678.

Bussey H (1975) Familial polyposis coli. Johns Hopkins University Press, Baltimore.

Bussey H (1978) Multiple adenomas and carcinomas. In B Morson (ed): The Pathogenesis of Colorectal Cancer: Vol 10. Major Problems in Pathology. WB Saunders, Philadelphia, pp 81–87.

Butt JH, Konishi F, Morson BC, Lennard-Jones JE, Ritchie JK (1983) Macroscopic lesions in dysplasia and carcinoma complicating ulcerative colitis. Dig Dis Sci 28:18–26.

Butt JH, Morson B (1981) Dysplasia and cancer in inflammatory bowel disease. Gastroenterology 880:865–868.

Cannon-Albright LA, Skolnick MH, Bishop DT, Lee RG, Burt RW (1988) Common inheritance of susceptibility to colonic adenomatous polyps and associated colorectal cancers. N Engl J Med 319:533–537.

Carcinogenesis and biochemical mechanisms: Animal models (1975) In JH Weisburger, BS Reddy, DL Joftes (eds): UICC Technical Report Series, Vol 19. International Union against Cancer, Geneva, pp 34–64. (Colo-Rectal Cancer)

Castleman B, Krickstein H (1962) Do adenomatous polyps of the colon become malignant? N Engl J Med 267:469–475.

Chang WWL (1978) Histogenesis of symmetrical 1,2-dimethylhydrazine-induced neoplasms of the colon in the mouse. J Natl Cancer Inst 60:1405–1418.

Chang WWL (1985) The mode of formation and progression of chemically induced

colonic carcinoma. In JRF Ingall, AJ Mastromarino (eds): Carcinoma of the Large
Bowel and Its Precursors. Alan R Liss, New York, pp 217–235.

Chapman I (1963) Adenomatous polypi of the large intestine: Incidence and distribution.
Ann Surg 157:223–226.

Christie J (1981) Comparative significance of right colon, left colon and rectal polyps.
Gastrointest Endosc 27:185–186.

Chu DZJ, Giacco G, Martin RG, Guinee VF (1986) The significance of synchronous carcinoma and polyps in the colon and rectum. Cancer 57:445–450.

Clark JC, Collan Y, Eide TN, Esteve J, Ewen S, Gibbs NM, Jensen OM (1985) Prevalence
of polyps in an autopsy series from areas with varying incidence of large-bowel cancer.
Int J Cancer 36:179–186.

Collins RH, Feldman M, Fordtran JS (1987) Colon cancer, dysplasia and surveillance in
patients with ulcerative colitis. N Engl J Med 316:1654–1658.

Cook MG, Goligher JC (1975) Carcinoma and epithelial dysplasia complicating ulcerative colitis. Gastroenterology 68:1127–1136.

Cooper HS, Farano P, Coapman RA (1987) Peanut lectin binding sites in colons of patients
with ulcerative colitis. Arch Pathol Lab Med 111:270–275.

Cooper HS, Marshall C, Ruggerio F, Steplewski Z (1987) Hyperplastic polyps of the colon
and rectum: An immunohistochemical study with monoclonal antibodies against blood
group antigens (Sialosyl-Le^a, Le^b, Le^x, Le^y, A, B, H). Lab Invest 57:421–428.

Cooper HS, Patchefsky AS, Marks G (1979) Adenomatous and carcinomatous changes
within hyperplastic colonic epithelium. Dis Colon Rectum 22:152–156.

Cooper HS, Reuter VE (1983) Peanut lectin–binding sites in polyps of the colon and rectum. Lab Invest 49:655–661.

Correa P, Strong JP, Johnson WD, Pizzolato P, Haenszel W (1982) Atherosclerosis and
polyps of the colon: Quantification of precursors of coronary heart disease and colon
cancer. J Chronic Dis 35:313–320.

Correa P, Strong JP, Reif A, Johnson W (1977) The epidemiology of colorectal polyps:
Prevalence in New Orleans and international comparisons. Cancer 39:2258–2264.

Craft CF, Mendelsohn G, Cooper HS, Yardley JH (1981) Colonic "precancer" in Crohn's
disease. Gastroenterology 880:578–584.

Crawford BE, Stromeyer FW (1983) Small nonpolypoid carcinomas of the large intestine.
Cancer 51:1760–1763.

Cuvelier CA, Morson BC, Roels HJ (1987) The DNA content in cancer and dysplasia in
chronic ulcerative colitis. Histopathology 11:927–939.

Deschner EE, Long FC, Maskens AP (1979) Relationship between dose time and tumor
yield in mouse dimethylhydrazine-induced colon tumorigenesis. Cancer Lett 8:23–28.

Diamond M (1972) Adenomas of the rectum and the sigmoid in alcoholics; a sigmoidoscopic study. Am J Dig Dis 19:47–50.

Dixon MF, Brown LJR, Gilmour HM, Price AB, Smeeton NC, Talbot IC, Williams GT
(1988) Observer variation in the assessment of dysplasia in ulcerative colitis.
Histopathology 13:385–398.

Dukes CE (1959) The etiology of cancer of the colon and rectum. Cancer 2:27–32.

Eddy DM, Nugent FW, Eddy JF, Coller JH, Gilbertsen V, Gottlieb LS, Rice R, Sherlock
P, Winawer S (1987) Screening for colorectal cancer in a high-risk population: Results
of a mathematical model. Gastroenterology 92:682–692.

Ehsanullah M, Naughton-Morgan N, Filipe MI, Gazzard B (1985) Sialomucins in the
assessment of dysplasia and cancer-risk patients with ulcerative colitis treated with
colectomy and ileorectal anastomosis. Histopathology 9:223–235.

Eide TJ (1983) Remnants of adenomas in colorectal carcinomas. Cancer 51:1866–1872.

Eide TJ (1986) Prevalence and morphological features of adenomas of the large intestine in individuals with and without colorectal carcinoma. Histopathology 10:111–118.

Eide T, Schweder T (1984) Clustering of adenomas in the large intestine. Gut 25:1262–1267.

Eide T, Stalsberg H (1978) Polyps of the large intestine in northern Norway. Cancer 42:2839–2848.

Ekelund GR, (1980) Cancer risk with single and multiple adenomas: Synchronous and metachronous tumors. In S Winawer, D Schottenfeld, P Sherlock (eds): Colorectal Cancer: Prevention, Epidemiology and Screening. Raven Press, New York, pp 151–155.

Ekelund GR, Lindstrom C (1974) Histopathological analysis of benign polyps in patients with carcinoma of the colon and rectum. Gut 15:654–663.

Figiel LS, Figiel SJ, Wietersen FK (1965) Roentgenologic observations of growth rates of colonic polyps and carcinoma. Acta Radiol [Diagn] (Stockh) 3:417–429.

Filipe MI (1969) Value of histochemical reactions for mucosubstances in the diagnosis of certain pathological conditions of the colon and rectum. Gut 10:577–586.

Filipe MI (1975) Mucous secretion in rat colonic mucosa during carcinogenesis induced by dimethylhydrazine – a morphological histochemical study. Br J Cancer 32:60–76.

Filipe MI, Branfoot AC (1976) Mucin histochemistry of the colon. Curr Top Pathol 63:143–178.

Fisher ER, Paulson JD, McCoy MM (1981) Genesis of 1,2-dimethylhydrazine-induced colon cancer. Arch Pathol Lab Med 105:29–37.

Fozard JBJ, Dixon MF, Axon ATR, Giles GR (1987) Lectin and mucin histochemistry as an aid to cancer surveillance in ulcerative colitis. Histopathology 11:385–394.

Franzin G, Novelli P (1982) Adenocarcinoma occurring in a hyperplastic (metaplastic) polyp of the colon. Endoscopy 14:28–30.

Franzin G, Zamboni G, Scarpa A, Dina R, Iannucci A, Novelli P (1984) Hyperplastic (metaplastic) polyps of the colon: A histologic and histochemical study. Am J Surg Pathol 8:687–698.

Giardiello FM, Welsh SB, Hamilton SR, Offerhaus GJA, Gittelsohn AM, Booker SV, Krush AJ, Yardley JH, Luk GD (1987) Increased risk of cancer in the Peutz-Jeghers syndrome. N Engl J Med 316:1511–1514.

Gilbertsen VA, Nelms JM (1978) The prevention of invasive cancer of the rectum. Cancer 41:1137–1139.

Gillespie PE, Chambers TJ, Chan KW, Doronzo F, Morson BC, Williams CB (1979) Colonic adenomas – a colonoscopic survey. Gut 20:240–245.

Goldgraber MB, Kirsner JB (1964) Carcinoma of the colon in ulcerative colitis. Cancer 17:657–665.

Goldin BR (1988) Chemical induction of colon tumors in animals: An overview. In G Steele, RW Burt, SJ Winawer, JP Karr (eds): Basic and Clinical Perspectives of Colorectal Polyps and Cancer. Alan R Liss, New York, pp 319–333.

Goodman ZD, Yardley JH, Milligan FD (1979) Pathogenesis of colonic polyps in multiple juvenile polyposis. Cancer 43:1906–1913.

Granqvist S (1981) Distribution of polyps in the large bowel in relation to age. A colonoscopic study. Scand J Gastroenterol 16:1025–1031.

Greenstein AJ, Sachar DB, Smith H, Janowitz HD, Aufses AH (1980) Patterns of neoplasia in Crohn's disease and ulcerative colitis. Cancer 46:403–407.

Greenstein AJ, Sachar DB, Smith H, Pucillo A, Papatestas AE, Kreel I, Geller SA, Janowitz HD, Aufses AH (1979) Cancer in universal and left sided ulcerative colitis: Factors determining risk. Gastroenterology 77:290–294.

Grinnell RS, Lane N (1958) Benign and malignant adenomatous polyps and papillary adenomas of the colon and rectum. An analysis of 1,856 tumors in 1,335 patients. Int Abstr Surg 106:519–538.

Grossman S, Milos ML, Tekawa IS, Jewell NP (1989) Colonoscopic screening of persons with suspected risk factors for colon cancer: II. Past history of colorectal neoplasms. Gastroenterology 96:299–306.

Grotsky HW, Rickert RR, Smith WD, Newsome JF (1982) Familial juvenile polyposis coli: A clinical and pathologic study of a large kindred. Gastroenterology 82: 494–501.

Gyde SN, Prior P, Macartney JC, Thompson H, Waterhouse JAH, Allan RN (1980) Crohn's disease and digestive tract cancer. Gut 21:1024–1029.

Hamilton SR (1985) Colorectal carcinoma in patients with Crohn's disease. Gastroenterology 89:398–407.

Hammarberg G, Slezak P, Tribukait B (1984) Early detection of malignancy in ulcerative colitis. A flow cytometric DNA study. Cancer 53:291–295.

Haviland AE, Borowitz MJ, Lan MS, Kaufman B, Khorrami A, Phelps PC, Metzgar RS (1988) Aberrant expression of monoclonal antibody–defined colonic mucosal antigens in inflammatory bowel disease. Gastroenterology 95:1302–1311.

Helwig EB (1947) The evolution of adenomas of the large intestine and their relation to carcinoma. Surg Gynecol Obstet Int Abstr Surg 84:36–49.

Hermanek P (1985) Diagnosis and therapy of cancerous adenoma of the large bowel: A German experience. In CM Fenoglio-Preiser, FP Rossini (eds): Adenomas and Adenomas Containing Carcinoma of the Large Bowel: Advances in Diagnosis and Therapy. Raven Press, New York, pp 57–62.

Hill M (1978) Etiology of the adenoma–carcinoma sequence. In B Morson (ed): The Pathogenesis of Colorectal Cancer. WB Saunders, Philadelphia, pp 158–166.

Hoff G, Clausen OP, Fjordvang H, Norheim A, Foerster A, Vatn MH (1985) Epidemiology of polyps in the rectum and sigmoid colon. Scand J Gastroenterol 20:983–989.

Hoff G, Moen IE, Trygg K, Frolich W, Savar J, Vatn M, Gjone E, Larsen S (1986) Epidemiology of polyps in the rectum and sigmoid colon: Evaluation of nutritional factors. Scand J Gastroenterol 21:199–204.

Hoff G, Vatn MH, Larsen S (1987) Relationship between tobacco smoking and colorectal polyps. Scand J Gastroenterol 22:13–16.

Horn RC, Payne WA, Fine G (1963) The Peutz-Jeghers syndrome. Arch Pathol 76:29–37.

Isaacson P, Attwood PRA (1979) Failure to demonstrate specificity of the morphological and histochemical changes in mucosa adjacent to colonic carcinoma (transitional mucosa). J Clin Pathol 32:214–218.

Isbister WH (1986) Colorectal polyps: An endoscopic experience. Aust NZ J Surg 56:717–722.

Jain M, Cook GM, Davis F, Grace MG, Howe GR, Miller AB (1980) A case-control study of diet and colorectal cancer. Int J Cancer 26:757–768.

Jarvinen H, Franssila, KO (1984) Familial juvenile polyposis coli; increased risk of colorectal cancer. Gut 25:792–800.

Jass JR (1983) Relation between metaplastic polyp and carcinoma of the colorectum. Lancet 1:28–30.

Jass JR, England J, Miller K (1986) Value of mucin histochemistry in follow up surveillance of patients with long standing ulcerative colitis. J Clin Pathol 39:393–398.

Jass JR, Filipe MI, Abbas S, Falcon CAJ, Wilson Y, Lovell D (1984) A morphologic and histochemical study of metaplastic polyps of the colorectum. Cancer 53:510–515.

Jass JR, Williams CB, Bussey HJR, Morson BC (1988) Juvenile polyposis—a precancerous condition. Histopathology 13:619–630.

Jones MA, Hebert JC, Trainer TD (1987) Juvenile polyp with intramucosal carcinoma. Arch Pathol Lab Med 111:200–201.

Kaneko M (1972) On pedunculated adenomatous polyps of the colon and rectum with particular reference to their malignant potential. Mt Sinai J Med NY 39:103–111.

Kjeldsberg CR, Altshuler JH (1970) Carcinoma in situ of the colon. Dis Colon Rectum 13:376–381.

Konishi F, Morson B (1982) Pathology of colorectal adenomas: A colonoscopic survey. J Clin Pathol 35:830–841.

Kozuka S, Nogaki M, Ozeki T, Masumori S (1975) Premalignancy of the mucosal polyp in the large intestine. II. Estimation of the periods required for malignant transformation of mucosal polyps. Dis Colon Rectum 18:494–500.

Kuramoto S, Oohara T (1988) Minute cancers arising de novo in the human large intestine. Cancer 61:829–834.

Labow SB, Hoexter B, Walrath DC (1979) Colonic adenocarcinomas in patients with ureterosigmoidostomies. Dis Colon Rectum 22:157–158.

LaMont JT, O'Gorman TA (1978) Experimental colon cancer. Gastroenterology 75:1157–1169.

Lane N (1976) The precursor tissue of ordinary large bowel cancer. Cancer Res 36:2669–2672.

Laqueur GL, Mickelsen O, Whiting MG, Kurland LT (1963) Carcinogenic properties of nuts from Cycas circinalis L. indigenous to Guam. J Natl Cancer Inst 31:919–951.

Lennard-Jones JE, Morson BC, Ritchie JK, Shove DC, Williams CB (1977) Cancer in colitis: Assessment of the individual risk by clinical and histological criteria. Gastroenterology 73:1280–1289.

Lennard-Jones JE, Morson BC, Ritchie JK, Williams CB (1983) Cancer surveillance in ulcerative colitis: experience over 15 years. Lancet 2:149–152.

Lescher TC, Dockerty MB, Jackman RJ, Beahrs OH (1967) Histopathology of the larger colonic polyp. Dis Colon Rectum 10:118–124.

Lev R (1979) On controlling colorectal cancer. Hum Pathol 10:621–623.

Lev R, Grover R (1981) Precursors of human colon carcinoma: A serial section study of colectomy specimens. Cancer 47:2007–2015.

Lev R, Herp A (1978) Pathogenesis of rat colon carcinomas induced by N-methyl-N-nitrosourea. J Natl Cancer Inst 61:779–786.

Lev R, Lance P, Camara P (1985) Histochemical and morphological studies of mucosa bordering recto-sigmoid carcinomas: Comparisons with normal, diseased and malignant colonic epithelium. Hum Pathol 16:151–161.

Linos DA, Dozois RR, Dahlin DC, Bartholomew LG (1981) Does Peutz-Jeghers syndrome predispose to gastrointestinal malignancy? Arch Surg 116:1182–1184.

Lipkin M (1986) The development of a risk profile for colorectal cancer by utilizing the proliferative and antigenic characteristics of colonic epithelial cells. Front Gastrointest Res 10:257–269.

Lipper S, Kahn LB, Sandler RS, Varma V (1981) Multiple juvenile polyposis: A study of the pathogenesis of juvenile polyps and their relationship to colonic adenomas. Hum Pathol 12:804–813.

Listinsky CM, Riddell RH (1981) Patterns of mucin secretion in neoplastic and non-neoplastic diseases of the colon. Hum Pathol 12:923–929.

Lotfi AM, Spencer RJ, Ilstrup DM, Melton LJ (1986) Colorectal polyps and the risk of subsequent carcinoma. Mayo Clin Proc 61:337–343.

Love RR, Morrissey JF (1984) Colonoscopy in asymptomatic individuals with a family history of colorectal cancer. Arch Intern Med 144:2209–2211.

Lovett E (1976) Family studies in cancer of the colon and rectum. Br J Surg 63:13–18.

Lynch HT, Albano WA, Ruma TA, Schmitz GD, Costello KA, Lynch JF (1983) Surveillance/management of an obligate gene carrier: The cancer family syndrome. Gastroenterology 84:404–408.

Lynch HT, Kimberling W, Albano WA, Lynch JF, Biscone K, Schuelke GS, Sandberg AA, Lipkin M, Deschner EE, Mikol YB, Elston RC, Bailey-Wilson JE, Danes BS (1985) Hereditary nonpolyposis colorectal cancer (Lynch syndromes I and II) 1. Clinical description of resource. Cancer 56:934–938.

Lynch HT, Watson P, Sanspa S, Marcus J, Smyrk T, Fitzgibbons R, Cristofaro G, Kriegler M, Lynch J (1988) Clinical nuances of Lynch syndromes I and II. In G Steele, RW Burt, SJ Winawer, JP Karr (eds): Basic and Clinical Perspectives of Colorectal Polyps and Cancer. Alan R Liss, New York, pp 177–188.

Macquart-Moulin G, Riboli E, Cornee J, Kaaks R, Berthezene P (1987) Colorectal polyps and diet: A case-control study in Marseilles. Int J Cancer 40:179–188.

Madara JL, Harte P, Deasy J, Ross D, Lahey S, Steele G (1983) Evidence for an adenoma–carcinoma sequence in dimethylhydrazine-induced neoplasms of rat intestinal epithelium. Am J Pathol 110:230–235.

Mannes GA, Maier A, Thieme C, Wiebecke B, Paumgartner G (1986) Relation between the frequency of colorectal adenoma and the serum cholesterol level. N Engl J Med 315:1634–1638.

Maratka Z, Nedbal J, Kocianova J, Havelka J, Kudrmann J, Hendl J (1985) Incidence of colorectal cancer in proctocolitis: A retrospective study of 959 cases over 40 years. Gut 26:43–49.

Maskens AP (1981) Confirmation of the two-step nature of chemical carcinogenesis in the rat colon adenocarcinoma model. Cancer Res 41:1240–1245.

Maskens AP (1983) Mechanisms of colorectal carcinogenesis in animal models: Possible implications in cancer prevention. In P Sherlock, BC Morson, L Barbara, U Veronesi (eds): Precancerous Lesions of the Gastrointestinal Tract. Raven Press, New York, pp 223–235.

Maskens AP, Dujardin-Loits RM (1981) Experimental adenomas and carcinomas of the large intestine behave as distinct entities: Most carcinomas arise de novo in flat mucosa. Cancer 47:81–89.

Matek W, Hermanek P, Demling P (1986) Is the adenoma–carcinoma sequence contraindicated by the differing location of colorectal adenomas and carcinomas? Endoscopy 18:17–19.

Mecklin J (1987) Frequency of hereditary colorectal carcinoma. Gastroenterology 93:1021–1025.

Ming-Chai C, Jen-Chun H, Pei-Yu C, Chi-Yuan C, Peng-Fei T, Shen-Hsing C, Fu-Pan W, Tsu-ling C, Shun-Chuan C (1965) Pathogenesis of carcinoma of the colon and rectum in Schistosomiasis japonica. A study of 90 cases. Chin med J [Engl] 84:513.

Morrison AS (1979) Sequential pathogenic components of rates. Am J Epidemiol 109:709-718.

Morrison AS (1985) Screening in Chronic Disease. Oxford University Press, New York.

Morson BC, Bussey HJR (1985) Magnitude of risk for cancer in patients with colorectal adenomas. Br J Surg [Suppl] 72:23-28.

Morson BC, Bussey HJR, Day DW, Hill MJ (1983) Adenomas of large bowel. Cancer Surv 2:451-478.

Morson BC, Pang LSC (1967) Rectal biopsy as an aid to cancer control in ulcerative colitis. Gut 8:423-434.

Muto T, Bussey HJR, Morson B (1975) The evolution of cancer of the colon and rectum. Cancer 36:2251-2270.

Muto T, Kamiya J, Sawada T, Kusama S, Itai Y, Ikenaga T, Yamashiro M, Hino Y, Yamaguchi S (1980) Colonoscopic polypectomy in diagnosis and treatment of early carcinoma of the large intestine. Dis Colon Rectum 23:68-75.

Neugut AI, Johnsen CM, Fink DJ (1986) Serum cholesterol levels in adenomatous polyps and cancer of the colon. A case-controlled study. JAMA 255:365-367.

Neugut AI, Johnsen CM, Forde KA, Treat MR, Nims C, Murray D (1988) Cholecystectomy and adenomatous polyps of the colon in women. Cancer 61:618-621.

Nugent FW, Haggitt RC, Colcher H, Kutteruf GC (1979) Malignant potential of colonic ulcerative colitis. Gastroenterology 76:1-5.

O'Higgins N, Digney J, Duff FA, Kelly DG (1981) Three polypoid colorectal tumors associated with different types of ureterocolic implantation. Br J Urol 53:278-279.

Perzin KH, Bridge MF (1982) Adenomatous and carcinomatous changes in hamartomatous polyps of the small intestine (Peutz-Jeghers syndrome). Cancer 49:971-983.

Petras RE, Mir-Madjlessi SH, Farmer RG (1987) Crohn's disease and intestinal carcinoma. A report of 11 cases with emphasis on associated epithelial dysplasia. Gastroenterology 93:1307-1314.

Peuchant E, Salles C, Jensen R (1987) Relationship between fecal neutral steroid concentrations and malignancy in colon cells. Cancer 60:994-999.

Pihl E, Peura A, Johnson WR, McDermott FT, Hughes ESR (1985) T-antigen expression by peanut agglutinin staining relates to mucosal dysplasia in ulcerative colitis. Dis Colon Rectum 28:11-17.

Ponz de Leon M, Roncucci L, DiDonato P, Tassi L, Smerieri O, Amorico MG, Malagoli G, DeMaria D, Antonioli A, Chahin NJ (1988) Pattern of epithelial cell proliferation in colorectal mucosa of normal subjects and of patients with adenomatous polyps or cancer of the large bowel. Cancer Res 48:4121-4126.

Potet F, Soullard J (1971) Polyps of the rectum and colon. Gut 12:468-482.

Pozharisski KM (1975) Morphology and morphogenesis of experimental epithelial tumors of the intestine. J Natl Cancer Inst 54:1115-1135.

Pozharisski KM, Chepick OF (1978) The oncological characteristics of colonic polyps in humans in view of morphogenesis of experimental intestinal tumors. Tumor Res 13:40-56.

Prager ED, Swinton NW, Young JL, Veidenheimer MC, Corman ML (1974) Follow-up study of patients with benign mucosal polyps discovered by proctosigmoidoscopy. Dis Colon Rectum 17:322-324.

Ransohoff DF, Riddell RH, Levin B (1985) Ulcerative colitis and colonic cancer: Problems in assessing the diagnostic usefulness of mucosal dysplasia. Dis Colon Rectum 28:383-388.

Reddy BS, Wynder EL (1977) Metabolic epidemiology of colon cancer: Fecal bile acids and neutral sterols in colon cancer patients and patients with adenomatous polyps. Cancer 38:2533–2539.

Restrepo C, Correa P, Duque E, Cuello C (1981) Polyps in a low-risk colonic cancer population in Columbia, South America. Dis Colon Rectum 24:29–36.

Rhodes JM, Black RR, Savage A (1986) Glycoprotein abnormalities in colonic carcinomata, adenomata and hyperplastic polyps shown by lectin peroxidase histochemistry. J Clin Pathol 39:1331–1334.

Richards WO, Webb WA, Morris SJ, Davis RC, McDaniel L, Jones L, Littauer S (1987) Patient management after endoscopic removal of the cancerous colon adenoma. Ann Surg 205:665–670.

Rickert RR, Auerbach O, Garfinkel L, Hammond EC, Frasca JM (1979) Adenomatous lesions of the large bowel. An autopsy survey. Cancer 43:1847–1857.

Riddell RH, Goldman H, Ransohoff DF, Appelman HD, Fenoglio CM, Haggitt RC, Ahren C, Correa P, Hamilton SR, Morson BC, Sommers SC, Yardley JH (1983) Dysplasia in inflammatory bowel disease: Standardized classification with provisional clinical applications. Hum Pathol 14:931–966.

Rider JA, Kirsner JB, Moeller HC, Palmer WL (1954) Polyps of the colon and rectum. Their incidence and relationship to carcinoma. Am J Med 16:555–564.

Ron E, Lubin F (1986) Epidemiology of colorectal cancer and its relevance to screening. Front Gastrointest Res 10:1–34.

Rosenstock E, Farmer RG, Petras R, Sivak MV, Rankin GB, Sullivan BH (1985) Surveillance for colonic carcinoma in ulcerative colitis. Gastroenterology 89:1342–1346.

Rotmensch S, Avigad I, Soffer E, Horowitz A, Bar-Meir S, Confino R, Czerniak A, Wolfstein I (1986) Carcinoma of the large bowel after a single massive dose of radiation in healthy teenagers. Cancer 57:728–731.

Sackett DL, Whelan G (1980) Cancer risk in ulcerative colitis: Scientific requirements for the study of prognosis. Gastroenterology 78:1632–1635.

Segal I, Cooke SAR, Hamilton DG, Tim LO (1981) Polyps and colorectal cancer in South African blacks. Gut 22:653–657.

Shamsuddin AM (1982) Microscopic intraepithelial neoplasia in large bowel mucosa. Hum Pathol 13:510–512.

Shamsuddin AKM, Trump BF (1981) Colon epithelium. II. In vivo studies of colon carcinogenesis. Light microscopic, histochemical, and ultrastructural studies of histogenesis of azoxymethane-induced colon carcinomas in Fischer 344 rats. J Natl Cancer Inst 66:389–401.

Shields HM, Bates ML, Goldman H, Zuckerman GR, Mills BA, Best CJ, Bair FA, Goran DA, DeSchryver-Kecskemeti K (1985) Scanning electron microscopic appearance of chronic ulcerative colitis with and without dysplasia. Gastroenterology 89:62–72.

Shinya H, Wolff WI (1979) Morphology, anatomic distribution, and cancer potential of colonic polyps. Ann Surg 190:679–683.

Simpson S, Traube J, Riddell RH (1981) The histologic appearance of dysplasia (precarcinomatous change) in Crohn's disease of the small and large intestine. Gastroenterology 81:492–501.

Skinner JM, Whitehead R (1981) Tumour-associated antigens in polyps and carcinomas of the human large bowel. Cancer 47:1241–1245.

Slater G, Fleshner P, Aufses AH (1988) Colorectal cancer location and synchronous adenomas. Am J Gastroenterol 83:832–836.

Smyrk T, Lynch H, Lanspa S, Kriegler M, Appelman H (1988) Adenomas in hereditary nonpolyposis colorectal cancer (HNPCC). Lab Invest 58:86A.

Spencer RJ, Melton LJ, Ready RL, Ilstrup DM (1984) Treatment of small colorectal polyps: A population-based study of the risk of subsequent carcinoma. Mayo Clin Proc 59:305–310.

Spjut HJ, Estrada RG (1977) The significance of epithelial polyps of the large bowel. Pathol Annu Pt I, 12:147–170.

Spjut HJ, Frankel NB, Appel MF (1979) The small carcinoma of the large bowel. Am J Surg Pathol 3:39–46.

Spjut HJ, Spratt JS (1965) Endemic and morphologic similarities existing between spontaneous colonic neoplasms in man and 3-2'-dimethyl-4-aminodiphenyl induced colonic neoplasms in rats. Ann Surg 161:309–324.

Spratt JS, Ackerman LV (1962) Small primary adenocarcinomas of the colon and rectum. JAMA 179:337–346.

Spratt JS, Ackerman LV, Moyer CA (1958) Relationship of polyps of the colon to colonic cancer. Ann Surg 148:682–698.

Stemmermann GN (1989) Geographic epidemiology of colorectal cancer: the role of dietary fat. In H Seitz, N Wright, V Simanaski (eds): Colorectal Carcinogenesis: An Approach to Cancer Prevention, Springer-Verlag, Heidelberg.

Stemmermann GN, Heilbrun LK, Nomura AMY (1988) Association of diet and other factors with adenomatous polyps of the large bowel: A progressive autopsy study. Am J Clin Nutr 47:312–317.

Stemmermann GN, Heilbrun LK, Nomura AMY, Yano K, Hayashi T (1986) Adenomatous polyps and atherosclerosis: An autopsy study. Int J Cancer 38:789–794.

Stemmermann GN, Yatani R (1973) Diverticulosis and polyps of the large intestine. A necropsy study of Hawaii Japanese. Cancer 31:1260–1270.

Stryker SJ, Wolff BG, Culp CE, Libbe SD, Ilstrup DM, MacCarthy RL (1987) Natural history of untreated colonic polyps. Gastroenterology 93:1009–1013.

Swaroop VS, Winawer SJ, Kurtz RC, Lipkin M (1987) Multiple primary malignant tumors. Gastroenterology 93:779–783.

Tanida N, Hikasa Y, Shimoyama T, Setchell KDR (1984) Comparison of faecal bile acid profiles between patients with adenomatous polyps of the large bowel and healthy subjects in Japan. Gut 25:824–832.

Terpstra OT, Blankenstein M, Dees J, Eilers GA (1987) Abnormal pattern of cell proliferation in the entire colonic mucosa of patients with colon adenoma or cancer. Gastroenterology 92:704–708.

Van der Werf SDJ, Nagengast FM, van Berge Henegouwen GP, Huijbregts AWM, Van Tongeren JHM (1982) Colonic absorption of secondary bile acids in patients with adenomatous polyps and in matched controls. Lancet 1:759–762.

Vatn MH, Stalsberg H (1982) The prevalence of polyps of the large intestine in Oslo: An autopsy study. Cancer 48:819–825.

Veale AMO (1965) Intestinal Polyposis. Eugenics Laboratory Memoirs, Series 40. Cambridge University Press, London.

Ward JM (1974) Morphogenesis of chemically induced neoplasms of the colon and small intestine in rats. Lab Invest 30:505–513.

Wargovich MJ, Medline A, Bruce WR (1983) Early histopathologic events to evolution of colon cancer in C57BL/6 and CF1 mice treated with 1,2-dimethylhydrazine. J Natl Cancer Inst 71:125–131.

Weedon DD, Shorter RG, Ilstrup DM, Huizenga KA, Taylor WA (1973) Crohn's disease and cancer. N Engl J Med 289:1099–1103.

Wiebecke B, Krey U, Lohrs U, Eder M (1973) Morphological and autoradiographical investigations on experimental carcinogenesis and polyp development in the intestinal tract of rats and mice. Virchows Arch [Pathol Anat] 360:179–193.

Williams AR, Balasooriya BAW, Day DW (1982) Polyps and cancer of the large bowel: A necropsy study in Liverpool. Gut 23:835–842.

Williams CB, Macrae FA (1986) The St. Mark's neoplastic polyp follow-up study. Front Gastrointest Res 10:226–242.

Williams GT, Arthur JF, Bussey HJR, Morson BC (1980) Metaplastic polyps and polyposis of the colorectum. Histopathology 4:155–170.

Winawer SJ, Ritchie MT, Diaz BJ, Gottlieb LS, Stewart ET, Zauber A, Herbert E, Bond J (1986) The National Polyp Study: Aims and organization. Front Gastrointest Res 10:216–225.

Winawer SJ, Zauber A, Diaz B, O'Brien M, Gottlieb LS, Sternberg SS, Waye JD, Shike M, National Polyp Study Work Group (1988) The National Polyp Study: Overview of program and preliminary report cf patient polyp characteristics. In G Steele, RW Burt, SJ Winawer, JP Karr (eds): Basic and Clinical Perspectives of Colorectal Polyps and Cancer. Alan R Liss, New York, pp 23–33.

Yardley JH, Bayless TM, Diamond TM (1979) Cancer in ulcerative colitis. Gastroenterology 76:221–224.

Yuan M, Itzkowitz SH, Ferrell LD, Fukushi Y, Palekar A, Hakomori S, Kim YS (1987) Expression of Lewis^x and sialylated Lewis^x antigens in human colorectal polyps. J Natl Cancer Inst 78:479–488.

5
Detection and Management

A. Screening, Diagnostic, and Surveillance Techniques*

In this section, there will be a discussion of the general principles of screening and their application to colorectal neoplasia. This will be followed by a description of the various methods of examining the colon, including for each technique the extent of accessible bowel, the sensitivity and specificity for detection of neoplastic lesions, and the complication rates of the procedures.

1. Principles of Screening

The identification and removal of polyps of the colon by screening may be a useful means of controlling cancer of the colon. *Screening for disease control* can be defined as the examination of asymptomatic people in order to classify them as likely, or unlikely, to have the disease that is the object of screening. People who appear likely to have the disease are investigated further to arrive at a final diagnosis. The organized application of early diagnosis and treatment activities in large groups often is described as "mass screening" or "population screening." The goal of screening is to reduce morbidity or mortality among the people screened by early treatment of the cases discovered.

To be suitable for control by a program of early detection and treatment, a disease must pass through a preclinical phase during which it is undiagnosed but detectable, and early treatment must offer some advantage over later treatment. Colorectal cancer does pass through a detectable preclinical phase, as evidenced by the detection of asymptomatic cases by screening (Winawer, 1979). Because of the presumed tendency of adenomatous polyps to progress to cancer, polyps also may be viewed as an early phase of colorectal cancer. Thus, polyps have been considered as a target of screening, along with cancer itself. Practical aspects of screening for polyps of the colon cannot be entirely separated from those of screening for colorectal cancer. The screening tests available detect both

---

*This section is derived in part from Morrison (1985).

types of lesions, and there may be uncertainty as to whether a polyp is benign or malignant until histological examination is done. Unanswered questions concern the relative importance of detecting and removing benign polyps versus early cancers in the control of colorectal cancer morbidity and mortality, the success with which polyps can be detected by tests available, and whether the high frequency of colorectal polyps is a deterrent to screening for them.

The immediate purpose of screening is to designate people with preclinical disease as "positive" and people without preclinical disease as "negative." Sensitivity and specificity measure the success of a screening test at performing these functions. Customarily, *sensitivity* is defined as the proportion of cases with a positive screening test among all cases of preclinical disease, as identified by a standard diagnostic test. Evaluating the sensitivity of a screening test (e.g., Hemoccult) can be problematic because of the difficulty or even impossibility of applying a diagnostic test (e.g., colonoscopy with biopsies as appropriate) to asymptomatic people with a negative screening test. Much of the available information on test sensitivity is derived from symptomatic persons. Test sensitivity appears to be substantially higher in symptomatic persons than in screened populations. A sensitive test, however, is a crucial aspect of any screening program, and practical measures of the sensitivity of screening tests are available. The *specificity* of a test is defined as the proportion of screened persons without preclinical disease who are correctly designated as negative. Satisfactory estimates are easier to obtain of specificity than of sensitivity. The specificity of a test is a major determinant of whether or not the frequency of false positives will be low enough for a screening program to be useful.

The predictive value (PV) in a screening program is the proportion of persons with a positive test who are found by diagnostic evaluation to have the disease in question. A high PV suggests that a reasonably high proportion of the costs of a program are in fact being expended for the detection of disease during the preclinical phase. A low PV suggests that a high proportion of the costs are being wasted on the detection and diagnostic evaluation of false positives, people who have a positive screening test but not the disease. The determinants of the PV of a given screening examination are the specificity and sensitivity of the test and the prevalence of preclinical disease in the target population. In practice, a low PV is more likely to be the result of poor specificity than of poor sensitivity. It is the specificity that determines the number of false positives. False positives are derived from persons without the disease, who constitute the vast majority of those tested in virtually any program. Although the proportion of false positives might be small, even a small loss of specificity can lead to a large increase in their number and a large decline in the PV.

The three major types of screening available for cancer, or polyps of the colon, are tests for blood in the stool, barium enema, and endoscopy. The apparent sensitivity, specificity, and predictive value of a screening test depend on the circumstances in which the measurements are made. The skill and training of the examiner, the laboratory techniques that are used, and the method of selecting subjects may all affect the observed values. Thus, the results given below are

intended as illustrations rather than as definitive descriptions of the characteristics of the tests.

2. The At-Risk Population to Be Screened

It should be stated at the outset that identification of such groups might be useful
for *investigative* screening studies only: Decreased mortality from colorectal
cancer has not yet been demonstrated in any screened groups to date. Populations
at risk for AP are generally the same as those at risk for CRC, although exceptions exist; for example, patients with inflammatory bowel disease are prone to
develop carcinoma but do not have a (documented) increased risk for AP. Subjects over 40 years of age, those with a previous history of AP or CRC, and those
with a family history of FPC clearly fall into this category. It is not known if subjects with a family history of hereditary nonpolyposis colorectal cancer (HNCC)
or of less clearly hereditary forms of CRC are at increased risk for APs (Burt et
al., 1985; Cannon-Albright et al., 1988; Grossman & Milos, 1988; see also Chap.
4.F.4). Thus, Cannon-Albright et al. (1988) found a statistically significant albeit
small increase in the prevalence of APs in relatives of subjects with CRCs and
APs as compared with spouse controls, whereas in their study of first-degree
relatives of subjects with CRC, Grossman and Milos (1988) found that adenoma
prevalence was no greater than in the normal population. The prevalence of APs
is increased in patients with breast carcinoma (Bremond et al., 1984), which is
of interest in view of the association between breast carcinoma and CRC (Lynch
II syndrome).

3. Digital Examination of the Rectum

The digital examination is used for carcinoma rather than for polyp detection.
Reliable data on the number of rectal polyps that are detectable digitally do not
exist. The American Cancer Society (ACS) has recommended that this examination be performed annually from the age of 40 onward. Depending on the patient
and the length of the examiner's finger, about 7.5 cm of bowel can be palpated on
digital examination, which will include the distal one half to two thirds of the rectum but no sigmoid colon. The examination is virtually without risk.

4. Stool Blood Tests

The stool blood test is recommended annually by the ACS starting from age 50.
The history and chemistry of testing for blood in the stool have been reviewed
by Fleisher et al. (1985). A number of tests have been developed. The majority
(e.g., Hemoccult) are based on the ability of hemoglobin-associated "pseudo-
peroxidase" to convert a colorless dye (e.g., gum guaiac) to a blue color in the
presence of H_2O_2. In the home-use Hemoccult test the patient prepares six slides
from three consecutive stools while on a high-residue (fiber), meat-free diet. A
newer technique (HemoQuant), based on the conversion of nonfluorescing heme

to fluorescent porphyrins, is more sensitive than Hemoccult, is highly reproducible, and gives quantitative information but is also more complex and expensive. Thus, Ahlquist et al. (1985) found that HemoQuant, using an upper limit of normal of 2 mg hemoglobin/g stool, was positive in 43 percent of polyps between 1 and 1.9 cm in size, whereas Hemoccult was positive in only 14 percent of the same subjects. Fleisher et al. (1985) and Simon (1985) discuss issues of quality control and of the diet of persons to be tested. The sensitivity of the fecal occult blood test for polyps of the colon has been found to vary with the location of the polyps, their size, and their histological features.

Useful data on the characteristics of testing for fecal blood are derived from the Memorial Sloan-Kettering Strang Clinic Study (Winawer, 1979). In this project, the "study" group was screened annually with both rigid sigmoidoscopy and the Hemoccult test. Therefore, the sensitivity of the latter test (for polyps in the rectosigmoid) can be evaluated in relation to sigmoidoscopy as a standard test. Of 21 patients who had a histologically confirmed rectosigmoid adenoma that was at least 5 mm in diameter, only 5 had a positive Hemoccult test. Thus, the estimated sensitivity of the Hemoccult test was $5/21 = 24$ percent. (In contrast, the sensitivity of the Hemoccult test for cancer was about 85 percent.)

Similar results using Hemoccult were reported by Bertario et al. (1988) in a study of 1,233 asymptomatic subjects. They found the sensitivity for 96 polyps to be 24 percent (95 percent confidence interval [CI]:16–33 percent) and for 20 carcinomas to be 75 percent (CI:54–91 percent). The sensitivity for APs was higher when they were multiple or were > 2 cm in size. Herzog et al. (1982) reported the results of the Haemoccult test (identical to the Hemoccult test) in patients with known polyps more than 15 mm in diameter. A three-day test was positive in 65 percent (22/34) with polyps of the descending colon or rectosigmoid but only in 30 percent (3/10) with polyps of the ascending or transverse colon.

Gabrielson et al. (1985) evaluated the Hemoccult II test with rehydration in patients referred for colonoscopy. A positive test was found in 26 percent (11/42) of patients with a polyp in the cecum or ascending colon, in 45 percent (19/42) of the patients with a polyp in the transverse or descending colon, in 50 percent (49/98) of patients with a polyp in the sigmoid but in only 18 percent (7/38) of patients with a rectal polyp. They also found that the relative frequency of positive Hemoccult II tests with rehydration increased from 20 percent (13/64) in patients whose polyps were less than 5 mm in diameter to 74 percent (20/27) in patients whose polyps were at least 20 mm in diameter. Similar results were reported by Macrae and St. John (1982). Gabrielson et al. (1985) also found a higher frequency of positivity for villous as compared with tubular adenomas and for pedunculated (versus sessile) and dysplastic (versus nondysplastic) polyps.

In the first two cycles of the Strang Clinic Study, 1.1 percent of those tested by Hemoccult had a positive result (Winawer, 1979). Of those for whom diagnostic results were available, 43 percent were found to have either colorectal cancer or an adenoma at least 5 mm in diameter. The remaining 57 percent were false positives, the numerator of the "nonspecificity." Since the number of screenees with negative tests is a good approximation of the number without neoplastic disease, the specificity of the Hemoccult test for cancer or polyps can be estimated as: $1 - (0.57 \times 0.011)/(1 - 0.011) = 0.994$, or 99.4 percent.

The predictive value of the Hemoccult test for adenomatous polyps (at least 5 mm in diameter) is the number of persons with confirmed polyps divided by the number of persons with positive tests. In the Strang Clinic Study, 24 persons with adenomatous polyps were found among 82 who had positive tests and then were fully evaluated. Thus, the predictive value was $24/82 = 29$ percent. (The predictive value for cancer was 13 percent.) The predictive value of the test for polyps is higher than the predictive value for cancer even though the sensitivity of the test is lower for polyps because the prevalence of polyps is much higher than the prevalence of cancer. The predictive value (cancer or polyps) increases strongly with age (Winawer et al., 1980). This trend reflects the increase in prevalence of colon neoplasia with advancing age.

The Hemoccult II test and the rehydrated test are both more sensitive than the ordinary Hemoccult I test. The increase in sensitivity is, however, achieved at a considerable loss of specificity and, therefore, predictive value. For example, the predictive value of Hemoccult II decreased from 44 percent to 19 percent with rehydration (Winawer et al., 1980). As a result, rehydration methods may not be suitable for screening (American Cancer Society, 1980; Fleisher et al., 1985).

5. Barium Enema (BE)

Suspensions of barium sulfate are used with the single or double contrast technique to identify structural abnormalities of the gastrointestinal tract. In single contrast studies, barium alone is used. For the double contrast technique, less barium is used and the walls of the bowel are then separated with gas, which is radiolucent. Double contrast studies are more sensitive than single contrast, with which polyps < 1 cm are likely to be missed. For polyp detection, only double contrast barium enemas (DCBEs) should be used.

The entire large bowel with the exception of the distal rectum is routinely visualized on a BE examination. Any patient having a BE should also have a sigmoidoscopy, and if the two are combined, all polyps of the large bowel are potentially accessible to detection. When carefully performed, DCBE is an accurate examination. In one consecutive series of 2,590 patients with a variety of colonic disorders, overall values of 84 percent, 84 percent, 93 percent, and 70 percent, respectively, were reported for the sensitivity, specificity, PV of a positive report, and PV of a negative report (Fork, 1983). Since BE has not been used for large-scale screening for colorectal neoplasia, however, useful data on its specificity and predictive value for polyps are not available. It has been used primarily as a confirmatory test in screening programs, so some information on its sensitivity for polyp detection can be obtained by comparing it with other confirmatory tests, especially colonoscopy, in positive screenees (see section A.6.d).

DCBE is a safe procedure. The main complication is perforation, which may be intra- or extraperitoneal. The frequencies for each are on the order of 1 in 12,000 and 40,000, respectively (Bartram & Kumar, 1981).

6. Endoscopy

The methods considered here are rigid sigmoidoscopy, flexible sigmoidoscopy, and colonoscopy. The most widely used method for large-scale screening used to be rigid sigmoidoscopy; a flexible instrument (35 or 60 cm) has largely replaced the rigid one in screening. The colonoscope is being used primarily for diagnosis.

a. Proctosigmoidoscopy

The three varieties of proctosigmoidoscope that are most widely used are the 25 cm rigid instrument and the 35 cm and 60 cm flexible endoscopes. For several reasons, the rigid proctosigmoidoscope should no longer be routinely used in the diagnosis of polyps. The procedure is more painful for the patient than the flexible instrument. The average depth of insertion is 20 cm or less, and the tip of the instrument is often not negotiated out of the rectum.

Currently, flexible proctosigmoidoscopy with the 35 cm instrument is recommended by the ACS at age 50 and, if negative, one year later and thereafter every three to five years for average-risk subjects. If these guidelines for polyp surveillance in asymptomatic subjects are to be implemented for the entire U.S. population, which may be impractical, the great majority of proctosigmoidoscopic examinations will have to be conducted by nonspecialists in the office or outpatient clinic. A consensus has evolved that in general the 35 cm flexible sigmoidoscope should be used by nonspecialists (Griffin, 1985). The 60 cm instrument is reserved for physicians with special training in fiberoptic gastrointestinal endoscopy. Data concerning the depth of insertion of either instrument should be interpreted with care. Penetration to 35 cm and 60 cm in the unstretched colon would place the tip of the endoscope in the distal descending colon and at the splenic flexure, respectively, in most cases. It is often implied that these anatomic landmarks indicate the extent of bowel routinely inspected with the short and long sigmoidoscopes. However, because of the elasticity of the colon, this is not the case. At a 60 cm examination the entire sigmoid colon is inspected in about 80 percent of patients (Lehman et al., 1983). Contrary to popular belief, insertion to 60 cm does not ensure that the splenic flexure has been reached. At a 30–35 cm examination, just over half the sigmoid colon is likely to be examined.

Most studies indicate that 40–50 percent of APs are located in the rectum and sigmoid colon (Schottenfeld & Winawer, 1982). Therefore, at best 50 percent, and perhaps only 25–35 percent, of APs are likely to be accessible, respectively, with the 60 cm and 35 cm sigmoidoscopes. Fecal contamination and location at flexures or behind folds hinder diagnosis, but in experienced hands most polyps > 0.5 cm in size should be detected. The number of diminutive polyps (< 0.5 cm) that go undiagnosed is unknown. Moreover, since many screening examinations are performed by occasional endoscopists, often in suboptimal circumstances, the diagnostic yield of APs of all sizes is likely to be less than that suggested in many published studies.

Complications

Sigmoidoscopy is a generally safe procedure. In one series of 16,325 procedures with the rigid instrument performed by the house staff, three perforations were reported (Nelson et al., 1982); others have reported even lower perforation rates (Befeler, 1967). In a series of 10,000 examinations performed by experienced endoscopists with the 60 cm instrument, 11 complications were noted (Winman et al., 1980). A separate survey of 17,167 procedures performed by primary-care physicians revealed four complications, including two perforations and two hemorrhages, all with the 60 cm sigmoidoscope; no complications were found in 5,296 procedures with the 35 cm instrument (Sanowski et al., 1985). Increased numbers of complications are encountered when biopsies are taken and polyps removed by electrocautery.

b. Colonoscopy

A considerable array of colonoscopes is now available, with working lengths from approximately 70 to 190 cm. Many gastroenterologists like to use an instrument that can be inserted to 170 cm in order to give themselves the best chance of reaching the cecum. Colonoscopy should be performed only by fully trained endoscopists. The experienced colonoscopist will be able to reach the cecum in 74–95 percent or more of patients, the percentage being higher in otherwise healthy subjects with a normal large bowel (Knutson & Max, 1979; Williams & Price, 1987). Evaluation of the anatomic level of insertion should be as objective as possible. However, as with sigmoidoscopy, the length of instrument that has passed the anal verge is often of little use for this purpose; having inserted the instrument to 120 cm, the tip may still be in the transverse colon. Anatomic appearances that are helpful include the circular haustral pattern in the descending colon and the triangular appearance of the folds in the transverse colon. The close proximity of the colon to the spleen and liver gives a bluish appearance at the splenic and hepatic flexures. Confirmation that the cecum has been reached comes from identification of the appendiceal orifice and the ileocecal valve and transillumination of the abdominal wall in the right iliac fossa. Fluoroscopy during colonoscopy is the most reliable method for confirming the anatomic level of insertion, but this facility is usually not routinely available in the endoscopy suite.

Complications

The major complications of colonoscopy are hemorrhage and perforation, but both are uncommon. In a survey of 6,614 colonoscopies performed by 674 gastroenterologists, compiled by the American Society for Gastrointestinal Endoscopy (ASGE) (1986), complication rates of 2.0 percent and 4.9 percent, respectively, were reported for patients undergoing diagnostic colonoscopy and colonoscopic polypectomy. The single death in the series followed surgery for a polypectomy-related perforation. The rates for hemorrhage and perforation were 1 percent and 0.1 percent in a series of 5,000 colonoscopies performed in a specialist unit in London, England (Macrae et al., 1983).

c. Sensitivity

When endoscopy has been used in *screening*, it has been viewed as the standard test. Thus, little information is available on the sensitivity of endoscopy. Williams and Macrae (1986) reported on the follow-up of 308 patients who had had an endoscopic polypectomy and were reexamined by rigid proctosigmoidoscopy, flexible sigmoidoscopy, colonoscopy, and double contrast barium enema. Some 74 patients were found to have new polyps by colonoscopy. Of these, 27 patients were found to have polyps by flexible sigmoidoscopy. Thus, the sensitivity of colonoscopy was 2.7 times (74/27) higher than that of flexible sigmoidoscopy in identifying patients with polyps. Rigid sigmoidoscopy has been found by numerous observers to be less sensitive than flexible sigmoidoscopy (Katon, 1979). In the series of 6,614 colonoscopies reported above (under "Complications"), and performed because of gastrointestinal bleeding, abnormal BE, or previous polypectomy (Hallstrom et al., 1984), polyps were found in 29 percent of cases. Although there is no way of knowing the sensitivity of polyp detection in this survey, it is reassuring that this figure is within the range reported for polyp prevalence from autopsy studies.

Almost no information on the specificity and predictive value of endoscopy is available. Implicitly, however, they must be very high (Macrae & Williams, 1985).

d. Comparisons with Barium Enema

The traditional response to the question of which diagnostic modality to use first for the detection of colorectal neoplasms is that endoscopy and barium radiology are complementary. The choice of investigation will to some extent be determined by the clinical setting. Many APs are discovered serendipitously in patients undergoing investigation for the irritable bowel syndrome, for abdominal pain, or for bleeding due to hemorrhoids. In practice, most authorities now recommend total colonoscopy as the initial investigation once a colorectal neoplasm is suspected. Colonoscopy has the advantage that biopsies can be taken and polypectomy performed as part of a diagnostic examination, whereas the identification of a lesion on DCBE usually necessitates a second, endoscopic procedure. In addition, colonoscopy is generally superior to DCBE for detecting polyps; sensitivity of colonoscopy for lesions > 0.5 cm has been estimated at 90 percent with a specificity of 100 percent, whereas comparable figures for DCBE are 60 percent and 95 percent, respectively (Barry et al., 1987). Chait et al. (1986) reported that only 47 percent of 45 adenomas > 0.5 cm in diameter were seen on DCBE. Williams and Macrae (1986) reported the sensitivity of DCBE to be 71 percent and of colonoscopy to be 92 percent in the detection of adenomas more than 0.7 cm in diameter. Diminutive polyps (< 0.5 cm) cannot be reliably detected by BEs of even the highest quality. Finally, the radiologist is at a disadvantage in cases of severe diverticular disease, when coexisting polypoid lesions are easily missed (Farrands et al., 1983).

Despite these remarks, colonoscopy is not invariably superior to DCBE. In one series of 600 patients examined by both modalities, 14 APs of > 0.7 cm seen on DCBE had been missed by the colonoscopist (Williams et al., 1982). Problem areas for the endoscopist include the hepatic and splenic flexures and, perhaps surprisingly, the rectum unless an inversion maneuver is performed.

7. Effectiveness of Screening

The most common method of evaluating the effectiveness of screening has been to compare the survival (or case fatality) of screen-detected cases with the survival of cases diagnosed because of symptoms. *Survival* is usually measured as the proportion of cases that survives a stated period after diagnosis, for example, five years. *Case fatality* is one minus the survival, or the proportion of cases that die within a given time after diagnosis. The apparent implication of longer survival among screen-detected cases is that early treatment leads to prolonged life. However, a comparison of the survival of screen-detected versus routinely diagnosed cases is impossible to interpret without additional information that usually is not available. Screen-detected cases may have higher survival than routinely diagnosed cases do for two reasons in addition to the effect of early treatment. First, patients whose disease is detected by screening gain lead time—the diagnosis of their disease is earlier than the time that clinical diagnosis would have occurred. Even if the time of death is unchanged, the proportion of cases that survives for some period after diagnosis (and the proportion of low-stage cases) will increase as a result of earlier diagnosis alone. Second, screening may identify cases destined to have a relatively benign course even if there is no lead time or reduction in mortality from early treatment. The "prognostic selection" would occur if screening programs attract volunteers who are relatively healthy and who will tend to have a favorable clinical course. It also is possible that screening itself preferentially identifies disease with a long preclinical phase ("length-biased sampling"). Presumably, patients with such disease would have a long clinical phase, that is, favorable survival. Therefore, comparisons of case fatality or survival are likely to suggest a benefit even if none exists and may greatly exaggerate the size of a true benefit as it would be reflected in the mortality rate of a screened population.

The ultimate gains derived from a screening program are reductions of serious illness and death among the people screened. For a study to be valid, the observed relation between the program and the outcome should reflect only the amount of gain derived from the program and not other differences between screened and unscreened groups. Therefore, the evaluation of screening must be on measures of disease occurrence that will not be affected by early diagnosis except to the extent that early treatment is beneficial. One such measure is the *overall mortality rate* – the rate of death from all causes. Other things being equal, the mortality rate will be lower in a screened than in an unscreened group only if early treatment prevents or postpones death from the disease screened for. However, the change in the overall mortality rate has an important drawback as a measure of efficacy. Frequently, the target disease accounts for only a small fraction of the deaths in a population. If this is the case, the overall mortality rate will be quite insensitive to a beneficial effect of screening. For example, even a marked reduction in colorectal cancer mortality would have only a small effect on the rate of mortality from all causes; a beneficial effect of screening might not be demonstrable by means of overall mortality rates. The disease-specific mortality rate (e.g., the colorectal cancer mortality rate) is a much more sensitive measure, and in most circumstances, it is the primary outcome to be assessed.

Two prospective experimental studies, which aim to evaluate screening for *fecal blood* in the reduction of mortality from colorectal cancer, have been in progress for several years. The Memorial Sloan-Kettering Strang Clinic Study compares annual screening by a test for fecal blood and by rigid sigmoidoscopy (the "study" group) with annual screening by sigmoidoscopy alone (the "control" group) (Winawer, 1979; Winawer et al., 1980). Thus, the primary purpose of the study is to assess the *additional* value of fecal blood testing in a program of sigmoidoscopic screening.

This is an intervention study, but the subjects are not assigned at random to one of the groups. Instead, the subjects are assigned according to the time period in which they attend the Strang Clinic. There were 13,127 persons assigned to the study group and 8,834 to the control group, all of whom were greater than 40 years of age. The Hemoccult test (six slides) was used annually at the start of the study; subsequently, the Hemoccult II test was used. The test was considered positive if any of the six slides were positive. Diagnostic examinations for confirmation of a positive screening test included colonoscopy and double contrast barium enema. Polyps that were found were removed and cancers treated.

This study was begun in 1974; definitive data on the effect of screening on colorectal cancer mortality have not yet been published. Preliminary results suggest a higher percentage of lower-stage carcinomas in the prevalence cases (those noted on the first screen) and a borderline improvement in CRC mortality among the Hemoccult-screened "initial" group of subjects, that is, those coming to the Strang Clinic for the first time, as compared with the controls (Flehinger et al., 1988). The actual data on survival differences between the initial study and control groups have not been presented, however, and as conceded by the authors, the extent to which lead time and length biases may have contributed to the differences is unknown.

The University of Minnesota Study compares groups screened with Hemoccult test yearly, every other year, or not at all (Gilbertsen, 1979; Gilbertsen et al., 1980). The study began in 1975 and will terminate in 1995. The period of recruitment ended in 1977. A total of 48,000 subjects who were at least 50 years old were assigned at random to one of the three groups. As in the Strang Clinic Study, six Hemoccult slides were used in each screen, and one or more positive slides was taken as a positive test. Near the end of the period of recruitment, rehydration of the slides before testing was adopted. Initially, both colonoscopy and single contrast barium enema were used as confirmatory tests. Later, the use of the barium enema was discontinued as a routine confirmatory test. As with the Strang Clinic Study, the mortality findings have not been published. Additional studies of fecal blood testing have begun recently, but results are not expected for several years (Chamberlain et al., 1986).

Two studies have addressed the value of *sigmoidoscopic screening*. Investigators at the Kaiser Permanente Medical Care Program carried out an experimental study of "multiphasic" screening (Dales et al., 1979; Friedman et al., 1986). Subjects were selected from among subscribers to the Kaiser Foundation Health Plan. Those eligible were 35–54 years old in 1964 and had been members of the plan for at least two years. A screened group of 5,156 individuals and a control group of 5,557 were designated by a random process. Members of the screened group were urged to have a "Multiphasic Health Checkup" annually. Members of the control group were not urged to have the checkups, but they could have them if they wished. As part of the checkup, a sigmoidoscopic examination was recommended for both men and women who were at least 40 years old. The process of urging members of the screened groups to have checkups continued until mid-1980. Some 84 percent of the screened group had had at least one multiphasic checkup, compared with 64 percent of the control group. Members of the screened group had had an average of 6.8 checkups per subject, and members of the control group had an average of 2.8. The authors compared the experience of the screened and control groups with respect to outpatient medical service utilization, frequency of hospitalization, disability, and mortality. Overall, the differences between the screened and control groups were small. Mortality from all causes combined was nearly identical for the screened and control groups. The 16-year cumulative mortality from colorectal cancer, however, was 2.3 per 1,000 in the screened group compared with 5.2 per 1,000 in the control group. This difference is intriguing, but the extent to which screening was responsible for it is unclear.

Dales et al. (1979) reported that sigmoidoscopy was done only about 1.5 times as often in the screened group as in the control group. Moreover, colorectal polyps were removed "at similar" frequencies in the two groups. There was a moderate tendency for colorectal cancer to be diagnosed at an earlier stage in the screened group than in the control group, but the cumulative incidence of the disease in the period from 1965 to 1974 was slightly greater in the control group (4.8 per 1,000) than in the screened group (4.1 per 1,000). Thus, it seems unlikely that the screening program advanced the diagnosis of either colorectal cancer or the presumed precursor lesion enough to lead to the 56 percent reduction in mortality suggested by the data. It is plausible that chance variability, or perhaps some unrecognized bias, was responsible for part or all of the mortality difference observed. Study data on lymphohematopoietic malignancy and suicide give some indirect support to this idea. The cumulative mortalities observed for these causes were in the opposite direction but similar in size to the values for colorectal cancer. Since the screening program almost certainly did not cause lymphohematopoietic malignancy or suicide, these differences must be regarded as consequences of unknown factors or chance.

Gilbertsen and Nelms (1978) described a nonexperimental follow-up study of sigmoidoscopic screening for benign neoplasms and cancer of the sigmoid colon and rectum. The incidence and mortality from rectosigmoid cancer in participants in a screening program consisting of annual physical examinations with proctosigmoidoscopy were compared with the incidence and mortality expected in a population of similar age and sex composition. There were 21,150 people who had at least one examination. These subjects were nearly equally divided between men and women at least 45 years old. Subsequently, this group had 92,650 additional screening examinations. Any benign lesions that were detected were removed, and cancers that were discovered were treated. Since a total of 113,800 examinations were done, participants had an average of 5.4 examinations.

Exact figures on follow-up were not given, but more than 92,000 person-years of experience were accumulated, approximately 4.4 years per subject. During the 25-year follow-up period, which began after the first screening examination, 13 rectosigmoid cancers were discovered by screening. Of these 13 cases, 1 died postoperatively, another died of what was, apparently, a primary cancer at another site, and the remaining cases were followed for at least five years without known recurrence. The authors estimated that about 90 cases of the sigmoid and rectum would be expected in 92,000 person-years of observation in a group with the age-sex composition of those participating in the screening program, compared with the 13 cases that were observed. Concerning the possibility of "volunteer" or "self-selection" bias in the screened group, it was stated that colonic cancers other than those of the sigmoid and rectum and extracolonic tumors occurred at about the same rates in screened and control groups. Although this report suggests that sigmoidoscopic screening is worthwhile, estimates of the value of annual screening cannot be made with much confidence from the results presented, even if the issue of self-selection is ignored.

There are two major questions to be answered. What effect does sigmoidoscopic removal of benign lesions have on the incidence of rectosigmoid cancer? And what is the effect of screening on rectosigmoid cancer mortality? In their comparison of the observed and expected incidence, Gilbertsen and Nelms excluded patients with cancers diagnosed at the initial screening examination. Some or all of these cases would have come to attention during the observation period if the initial examination had not been done. Twenty-seven cases were found at the initial examination of the 21,150 participants. Thus, as many as 40 cases, rather than 13, might have come to attention if those discovered at the initial screen are included. Furthermore, potential subjects were asked about gastrointestinal symptoms, and only those who stated that they had none could become participants. Obviously, no such restriction could be placed on the "general population" comparison figures, so that participants in the program would be expected to be at comparatively low risk even if no screening were done. Similar problems affect the interpretation of the figures given on deaths. The postoperative death in one of the cases detected at a follow-up screen must be attributed to cancer. Of the first 25 cases detected at the initial examination, 9 died within five years of diagnosis. Assuming that these deaths are all attributable to rectosigmoid cancer, the rectosigmoid cancer mortality rate in the study would be roughly $(9 + 1)/92,000$ person-years $= 1$ per 10^4 person-years. This rate may be lower than that in the general population with the same age and sex distribution, but the general population mortality rate is not a suitable comparison figure. Since people known to have cancer were excluded from screening, the correct comparison figure would be the death rate from sigmoid and rectal cancer in the people who were free of clinical disease at the start of observation. This figure could be derived from incidence and survival data, although, again, there is no obvious way to make an adjustment for the fact that program participants had reported being free of symptoms at the start. Another objection to the Gilbertsen and Nelms study has been voiced by Miller (1988), who disputed their estimate of 90 cases of expected rectosigmoid carcinomas in the screened group. Miller's calculations, based on SEER (surveillance, epidemiology, and end results) incidence data, are that only 38 carcinomas would be expected and that this figure is very close to the sum of the 13 found by Gilbertsen and Nelms during the follow-up period added to the 27 carcinomas they detected on initial examination.

To *summarize*, definitive results are not available from experimental studies of fecal blood testing. A substantial difference in colorectal cancer mortality between screened and control groups has not yet been found (Simon, 1985; Chamberlain et al., 1986), although screening does increase the detection of APs, especially larger ones, and of CRCs and the proportion of CRCs that are detected as "early" lesions (see Hardcastle et al., 1986). After reviewing the available data, the United States Preventive Services Task Force found insufficient evidence to recommend either for or against fecal occult blood screening for people > 45 years (Knight et al., 1989), despite current ACS policy, which favors such testing. An experimental study focusing primarily on the efficacy of sigmoidoscopy has not yet been done. The sigmoidoscopic studies that have been performed do suggest, again, that APs and early CRCs may be detected using that modality, but the effects of repeated sigmoidoscopies/polypectomies on CRC mortality are open to serious questions of interpretation. Thus, there is, as yet, no convincing evidence that sigmoidoscopic screening with the removal of detected polyps and asymptomatic cancers leads to reduced mortality from colorectal cancer (Greenwald & Sondik, 1986). Similar conclusions were reached by Miller (1988) and by Neugut and Pita (1988), who agreed that data on the benefits of screening sigmoidoscopy were too meager to justify its widespread usage. It should also be pointed out that although the American Cancer Society supports such screening, other major organizations such as the Canadian Task Force on Periodic Health Examinations and the United States Preventive Services Task Force (Selby & Friedman, 1989) do not, at least in average-risk people > 40 years of age.

There is disagreement as to the significance of the finding of low-stage carcinomas by the above screening modalities. Since the prognosis of low-stage carcinomas is superior to that of higher-grade tumors, their detection would appear to impart a survival advantage to subjects bearing them, and in fact, detection of low-stage carcinomas has been described as "an acceptable analogue for mortality reduction" (DeCosse, 1988). Others claim that this overlooks length bias, which suggests that screening simply detects more slowly growing and more biologically favorable tumors, and lead-time bias, meaning that although the interval between diagnosis and death is prolonged, the time of death is unchanged. Further consideration of this issue is beyond the scope of this book.

B. The Malignant Polyp

1. The Problem

One unique aspect of management of patients undergoing polypectomy involves the malignant polyp, that is, an AP demonstrating malignant transformation with invasion of the submucosa. Polypoid carcinomas with or without residual adenomatous tissue may be included in this category, but most workers exclude lesions extending into the muscularis propria or beyond even when they have a polypoid or papillary surface: The term *malignant polyp* should be confined to a lesion that is potentially removable by endoscopic polypectomy. The main issue to be considered here is whether such polypectomy is sufficient or whether some form of subsequent colonic resection is required. This will depend on the likelihood of residual carcinoma being present in the bowel wall or in the regional (or, more rarely, distant) lymph nodes.

2. Frequency

The frequency of malignancy in a polyp will vary according to the population under study. In surgical series, which often contain higher numbers of larger polyps, it ranges between 5 percent and 11 percent (Grinnell & Lane, 1958; Day & Morson, 1978; Muto et al., 1980). A frequency of 9.2 percent was recorded for the Erlangen Polyp Registry, which included, as of 1982, 4,715 APs (Hermanek, 1985). In typical colonoscopic studies, it is in the 2–3 percent range (Frühmorgen & Matek, 1983; Isbister, 1986; Wegener, 1986) and in autopsy series about 1 percent (e.g., Rickert et al., 1979).

3. Risk Factors for Malignancy in a Polyp

These factors, which have already been alluded to in Chap. 2.B and 2.C, include the following:

TABLE 5.1. Relationship between degree of villosity and frequency of malignancy in polyps.

Type of polyp	Percentage of all polyps	Percentage malignant
Tubular adenoma	75.0	4.8
Tubulovillous adenoma	15.3	22.5
Villous adenoma	9.7	40.7

SOURCE: Modified from tables 5 and 7 in Muto et al., 1975.

a. Size

Polyps < 1.0 cm in diameter have about a 1 percent chance of being malignant; those between 1.0 and 2.0 cm, a 5–10 percent chance; and those > 2.0 cm, between a 20 and 48 percent chance.

b. Multiplicity

The chance that a patient has or will develop invasive carcinoma is greater in the presence of multiple polyps (Muto et al., 1975). It is assumed that such carcinomas develop from one of the polyps, as in FPC, but the evidence here is more often circumstantial rather than histopathological. The link between multiple polyps and carcinoma is further strengthened by the fact that with increasing numbers of APs a greater percentage of them show severe dysplasia (Shinya & Wolff, 1979; Morson et al., 1983).

c. Villous Component

The probability of malignancy in a polyp rises with the percentage of villosity in that polyp. Typical figures are shown in Table 5.1. Some workers feel that purely tubular adenomas never show carcinoma and that cases reported as such in the literature are really tubular adenomas with minor villous components (Spjut & Estrada, 1977; Spjut & Appel, 1979).

d. Degree of Dysplasia

A good correlation has been noted between the degree of dysplasia and the presence of carcinoma in a polyp (Table 5.2). It has been postulated that in the few malignant polyps showing only mild dysplasia elsewhere in the polyp, inadequate sampling may have missed the more severely dysplastic areas. Although the degree of dysplasia is generally proportional to polyp size, severe dysplasia may be found in small polyps and undoubtedly accounts for the occasional carcinomas found in polyps < 1.0 cm in diameter.

4. Effects of Study Design and Polyp Processing Technique

In some studies, primary colectomy was performed and the contained polyp subsequently removed via polypectomy and analyzed (e.g., Colacchio et al., 1981;

TABLE 5.2. Relationship between degree of dysplasia and frequency of malignancy in polyps.

Degree of dysplasia	Number of polyps (% of total)	Number with malignancy (%)
Mild	1,734 (69.2)	99 (5.7)
Moderate	549 (21.9)	99 (18.0)
Severe	223 (8.9)	77 (34.5)

SOURCE: Modified from table 9 in Muto et al., 1975.

Haggitt et al., 1985). This does not conform to the usual situation in which a decision regarding subsequent resection must be based on results from initial polypectomy. Another confounding factor is the inclusion in some studies of cases of polyps containing only carcinoma in situ (Muto et al., 1980; Haggitt et al., 1985; Richards et al., 1987); these are not relevant to the management of malignant polyps, which by definition show invasive carcinoma, since the in situ lesions have no metastatic potential. Finally, some studies include lesions that are invasive into muscularis propria or beyond (Coustifides et al., 1979).

The technique of processing the specimen is critical to the interpretation of the results (Morson et al., 1984; Riddell, 1985; Wilcox et al., 1986). Whenever possible, the polyp should be removed intact rather than piecemeal. The polyp must then be fixed such that perpendicular sections can be obtained through its center in continuity with the underlying stalk and resection margin. To ensure this, it has been recommended that a pin be placed through the polyp and stalk prior to fixation. The resection margin, recognized by cautery artifact, may have to be submitted separately if the polyp is too large and/or the stalk too long. Coating the margin with India ink may further help in its identification in tissue sections. Even under these ideal circumstances, it is not always possible to demonstrate definite penetration by carcinoma into submucosa, especially in the presence of splaying and duplication of the fibers of the muscularis mucosae, which commonly occur in the heads of polyps (see Figures 5.1 and 5.2 and Lipper et al., 1983).

There is disagreement in the literature regarding margin invasion. Some consider the margin to be involved when carcinoma extends to within 1 mm of the margin (Lipper et al., 1983), whereas others use 2 mm as a cutoff (Cranley et al., 1986).

Adequate numbers of sections of the polyp are necessary to provide reliable results. Hermanek (1985) has calculated an error of 14 percent, including wrong diagnoses or unassessable results, when a single section through the center of the polyp is submitted. He also found an error of 6 percent when 420 μm step sections are submitted and therefore recommends that 200 μm step sections be taken.

Should subsequent colonic resection be anticipated, it has been recommended that the polypectomy site be tattooed with a submucosal injection of 1-2 ml of autoclaved India ink (Williams & Whiteway, 1985); the tattooed area will be visible as a blue gray stain from the serosal and mucosal aspects for many weeks.

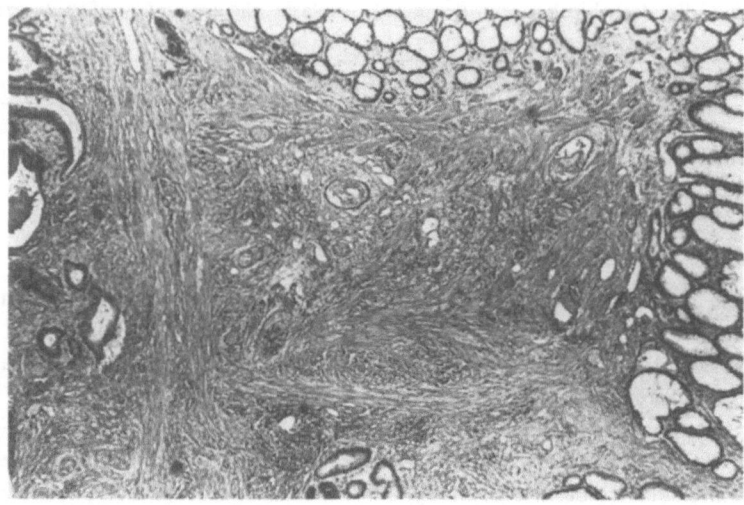

FIGURE 5.1. Head of benign adenomatous polyp showing extensive proliferation of smooth muscle fibers. Normal mucosa is noted above and to the right; adenomatous glands, some surrounded by muscle tissue, line the opposite margins. H&E; ×22, reproduced at 100%.

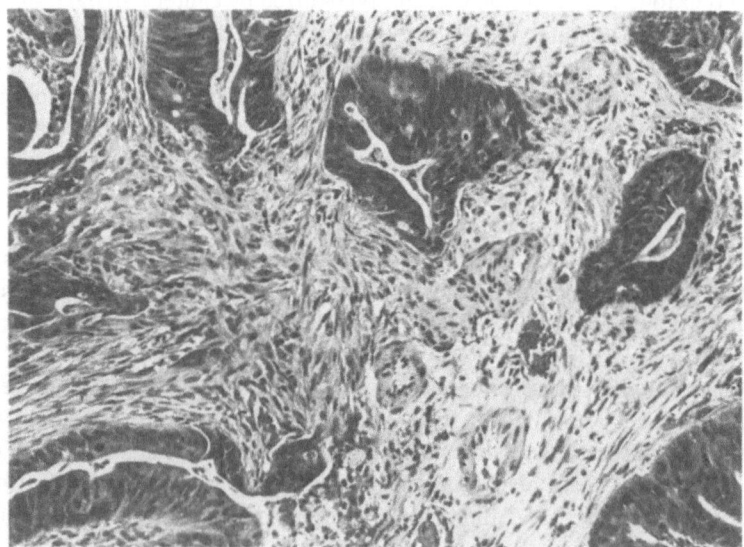

FIGURE 5.2. From short-stalked malignant polyp. Carcinomatous glands are embedded in a stroma containing proliferating muscle fibers, fibroblasts, and arterioles. It may be difficult to determine if such glands are in lamina propria or submucosa. H&E; ×176, reproduced at 55%.

Without this treatment the site will be unidentifiable within several weeks after polypectomy.

5. Statistical Analysis

This aspect has been examined critically by Wilcox and colleagues (Wilcox et al., 1986; Wilcox & Beck, 1987). They have developed a "decision tree" for determining the probability of success of a postpolypectomy resection, factoring into the equation

1. The chance of residual carcinoma in the bowel wall or lymph nodes of the specimen, which they estimated at 4.5 percent;
2. The operative efficacy or the percentage of patients with residual disease who are curable by resection; the rest are failures owing to the presence of unresectable residual disease or distant metastases; and
3. The operative risk, estimated at 0.2 to 2.0 percent.

They have concluded that it would be necessary to obtain 148 cases without an *adverse outcome* (absence of residual disease in bowel wall or lymph nodes at time of resection or during extended follow-up period) to justify at 95 percent confidence levels no further surgery, assuming an operative mortality risk ≤ 2 percent. They pointed out that most previous studies involved small numbers of cases and/or wide confidence intervals. Even the study of Morson et al. (1984), in which 0/49 polypectomized patients had adverse outcomes on follow-up, could not exclude a 7 percent risk of adverse outcome at the 95 percent confidence level. That risk is much higher than the operative mortality risk of 2 percent used by Wilcox et al. in their calculations.

There is some disagreement as to the length of follow-up period required to exclude an adverse outcome. Some workers feel that 5 years is adequate, whereas others prefer to wait 10 years, mainly to rule out lead-time bias.

6. Adverse Outcomes

Adverse outcomes, as defined above, occur on the average in about 10 percent of the cases, although figures up to 25 percent have been recorded; the exact number will partly depend on how cases are selected for polypectomy. This risk must be balanced against the risk of operative mortality of colon resection, as discussed above by Wilcox et al. and by others. The risk factors that will determine the probability of an adverse outcome in a malignant polyp are

1. Inadequacy of excision of the polyp as evaluated grossly by the endoscopist,
2. Inadequacy of excision as judged microscopically by inspecting the cauterized margin of resection,
3. Involvement of the polyp stalk or bowel wall submucosa by carcinoma as determined histologically,

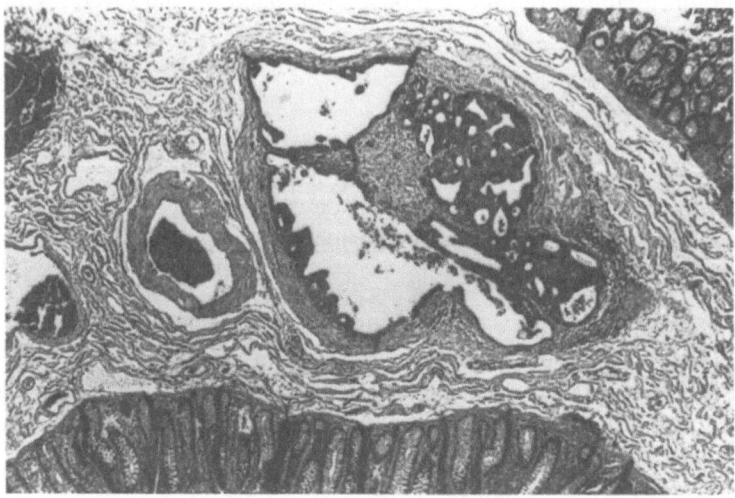

FIGURE 5.3. Stalk of malignant polyp. Clusters of tumor are noted in a large vein but not in the adjacent artery. Normal mucosa is seen below and in upper right. H&E; ×22, reproduced at 100%.

4. Poor histological differentiation (grade III) of the carcinoma, and
5. The presence of carcinoma in lymphatic or venous channels of the stalk (Figure 5.3).

The majority of investigators do not feel that sessile configuration, presence of a short rather than a long stalk (Cooper, 1983), polypoid carcinoma without adenomatous remnants, and villous histology constitute risk factors per se independent of the five listed above. They also believe that adverse outcomes occur only in the presence of one or more of those five factors. Colacchio et al. (1981), however, found that of six subjects with regional lymph node metastases, two had none of these risk factors in the excised polyp. They concluded that this represented too high a number of unpredictable adverse outcomes and recommended that all patients demonstrating malignancy in a polyp at polypectomy should be subjected to colon resection, a belief that has been incorporated into a recent textbook on the subject (Steele & Osteen, 1986). This view has few adherents. Most workers do not find nearly so high a rate of unpredictable adverse outcomes and feel that if care is exercised in selecting cases for endoscopic polypectomy, the cited risk factors are invaluable in defining the small subset of patients requiring surgical resection.

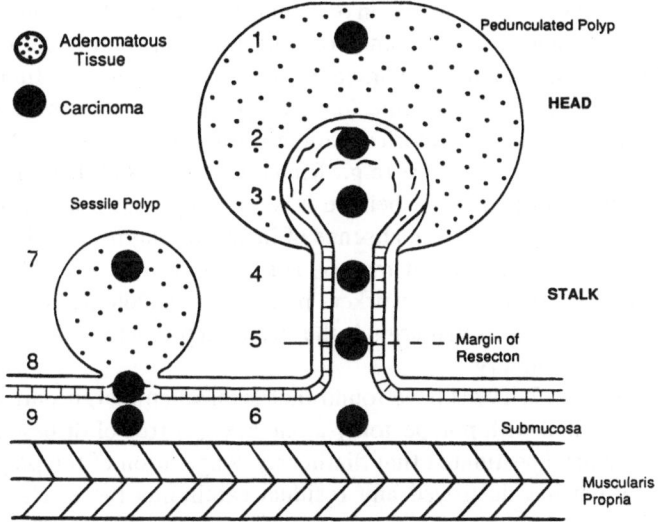

FIGURE 5.4. Diagram showing carcinomatous invasion at various levels within pedunculated and sessile malignant polyps. *1 and 7,* Carcinoma in situ; no metastatic potential; not a true malignant polyp; in 7, it may be unclear if malignant glands are in mucosa or submucosa if stromal response present or if section orientation is nonoptimal; *2 and 8,* malignant glands admixed with fibers of muscularis mucosae; it may be difficult to exclude superficial submucosal invasion; *3,* carcinoma in submucosa of head of polyp; very rarely associated with nodal metastases; *4,* carcinoma in submucosa of stalk; *5,* carcinoma at resection margin; *6 and 9,* carcinoma in colon wall submucosa; rarely demonstrable in polypectomy specimens.

7. Level of Submucosal Invasion by Carcinoma

The level of submucosal invasion by carcinoma is the single most important factor determining the possibility of an adverse outcome. The diagram in Figure 5.4 depicts sites of carcinoma in polyps and lists some potential pitfalls in the histopathological interpretation of the depth of invasion. True submucosal infiltration by carcinoma must be differentiated from pseudocarcinomatous invasion, which is found in about 2.5 percent of APs. Such islands of entrapped glands are often accompanied by mucosal lamina propria and by fresh hemorrhage and hemosiderin pigment indicative of earlier hemorrhage. They do not usually show either the severe dysplasia or the stromal reaction associated with invasive carcinoma. The condition is believed to result from repeated trauma (torsion, hemorrhage, erosion) to pedunculated polyps with herniation of adenomatous tissue into submucosa through the damaged muscularis mucosae. It can generally be recognized by experienced pathologists.

Most investigators feel that carcinoma confined to the submucosa in the head of the polyp does not warrant colectomy (e.g., Christie, 1984) in view of the rarity of metastases from such a lesion (Grinnell & Lane, 1958). Opinion is more evenly split regarding the significance of stalk invasion. Some feel that in the absence of resection margin involvement or other risk factors this finding is *not* associated with an adverse outcome (e.g., Lipper et al., 1983; Morson et al., 1984; Haggitt et al., 1985; American Society for Gastrointestinal Endoscopy Guidelines, 1986), whereas others (e.g., Richards et al., 1987) feel that it is. My impression is that most surgeons believe it is too dangerous not to resect when there is stalk invasion even in the absence of involved margins or other risk factors, although surgical sentiment may be shifting toward a more conservative approach (Christie, 1988). Most workers believe that histologically demonstrable carcinoma at the resection margin is the strongest predictor of an adverse outcome and warrants surgery.

Morson et al. (1977, 1984) have found that even when margins are involved, most subjects treated with polypectomy alone are still free of disease after five years, and they have postulated that diathermy coagulation of the polypectomy site in the patient has destroyed any residual carcinoma there. Haggitt et al. (1985) have emphasized the high sensitivity of level IV (bowel wall submucosa) invasion by carcinoma, but only rarely do endoscopic polypectomy specimens include substantial amounts of that tissue. It is uncertain, parenthetically, why tumor in bowel wall submucosa has so much worse a prognosis than in polyp head submucosa; the latter clearly contains lymphatics and good-sized arterioles and venules. The difference may be due to the greater number of available vascular channels in bowel wall submucosa or their greater size. Vessel size may be significant in view of the demonstration that prognosis of frank rectal carcinoma is worse when the tumor involves large, thick-walled (extramural) veins rather than thin-walled veins (Talbot et al., 1981). On the other hand, it has been shown by at least one group (Minsky et al., 1988) that tumor in (smaller) lymphatics is associated with decreased survival (again, in frank carcinoma), so that vessel-size factor must still be considered controversial.

In malignant polyps, undifferentiated histology and lymphatic permeation by carcinoma, while having high predictive values, are uncommon and thus have low sensitivities and less usefulness than resection margin involvement.

Insufficient attention is being paid to tumor invasion of vessels in the stalk. It may be difficult to differentiate tumor in vascular spaces from retraction artifact and to differentiate blood vascular from lymphatic channels. In addition to elastic stains, which help identify larger vessels, electron microscopy and some of the newer endothelial and basement membrane stains—for example, factor VIII and laminin—may be useful in this regard.

8. Postpolypectomy Therapy

Should a conservative course be decided upon, endoscopic examination of the polypectomy site 4–8 weeks later to rule out residual disease and a second "late" colonoscopy at 6–12 months after polypectomy are recommended by

Christie (1984). A detailed discussion of postpolypectomy surveillance is given in section D.

Should surgical resection be necessary, preliminary staging with computerized tomography scan and lymph node imaging may be of use in justifying more conservative surgery, especially for malignant rectal polyps that might have required abdominoperineal resection (DeCosse, 1984; see also Heberer et al., 1978).

C. Endoscopic Polypectomy

1. How Polyps Are Detected

There are three ways in which APs come to be diagnosed. First, and most often, polyps are discovered accidentally in the course of investigations for conditions such as colonic dysfunction (the irritable bowel syndrome) or rectal bleeding that they did not cause. Second, they are detected in asymptomatic subjects who are enrolled in screening programs for early diagnosis of CRC. Third, and least often, a large polyp may be the cause of symptoms for which investigations were undertaken.

2. Rationale for Detection and Eradication of Polyps

Once diagnosed, the currently accepted dogma is that APs must be extirpated. However, from a public health standpoint the degree of effort that should be expended on detecting asymptomatic APs is far from clear. The components of screening programs for CRC are digital examination of the rectum, fecal occult blood testing (FOBT), and proctosigmoidoscopy (Eddy, 1984). Most of the data that are used to justify screening come from FOBT programs. The declared objective is to detect early invasive CRC (Dukes stage A), for which curative resection is possible in > 80 percent of cases compared with little better than 40 percent overall for symptomatic CRC (Olson et al., 1980; Wilson, 1986). Implicit also is the intention to detect and remove as many adenomas as possible, since the majority of CRCs are thought to arise from such APs.

Simple and attractive as this approach may seem, the costs and benefits of adenoma detection in asymptomatic subjects demand critical examination. The majority of adenomas that are now extirpated endoscopically were never destined to become cancers. Although some large adenomas and early CRCs are detected in such asymptomatic individuals, there is as yet no direct evidence that FOBT or sigmoidoscopic screening programs or excision of APs leads to reduced mortality from CRC, as discussed in section A above. In addition, no reliable data are available on such factors as the length of time for which a precancerous adenoma is detectable before it becomes invasive, the proportion of precancerous adenomas that bleed before they become invasive, and the length of time for which bleeding is detectable prior to invasion. Histological and topographical features of polyps and carcinomas that may account for bleeding have been analyzed (Sobin 1985), but these features have not been correlated with the above clinical factors. Thus, scientifically rigorous studies of screening for colorectal neoplasms are urgently required to validate the current screening recommendations. In one such study, a mathematical model was constructed to help determine the frequency of rigid sigmoidoscopy necessary to detect curable neoplasms in asymptomatic subjects (Carroll & Klein, 1980). They factored into the equation the estimated doubling time of the polyp beyond the time of its detectability (5 mm) such that the polyp size would not exceed 13 mm. They estimated that sigmoidoscopic examinations every two to three years would be sufficient to detect the majority of polyps, even the fairly aggressive ones, before they grew to this size and thereby acquired significant malignant potential.

3. Manpower and Cost Considerations

There are approximately 30 million people in the United States walking around with tiny polyps of the colon. If the colonoscopists attempt to remove all of these, it will cost more than the national debt. (Marshak, 1981, p. 628)

The human and financial costs of screening programs are considerable (Frank, 1985; Simon, 1985). The morbidity and mortality of diagnostic and therapeutic colonoscopy have already been discussed. Since the positive predictive value of FOBT for colorectal neoplasms is 50 percent or less, many screened subjects devoid of neoplasia will undergo unnecessary colonoscopy. It has been estimated that 25 percent of compliant individuals with a normal life span, who are enrolled in screening programs, will require colonoscopy in due course for a positive FOBT, regardless of whether they have an AP or carcinoma.

Published data concerning the expense of screening programs are scanty. The cost of investigating a subject with a positive FOBT in New Mexico in 1980 was calculated as $700 (Applegate & Spector, 1981), which comes to about $11 per subject in the program. In 1990, over 70 million Americans will be > 45 years of age, the recommended starting age for screening programs. Assuming that the current costs have doubled and that 2 percent of those screened will have a positive FOBT, it will cost about $1.4 billion annually to work up the positive screenees. This expenditure might be justified if the mortality and cost of treating CRC are reduced accordingly, but evidence to that effect does not yet exist.

Along with the cost, an enormous expansion in colonoscopy facilities will be necessary if populationwide FOBT testing is demanded by all those for whom it is currently recommended. With existing positive rates, about 1 million colonoscopies would have to be performed annually in North America (Frank, 1985). Of course, patient compliance is unlikely to be such that this will be necessary.

Some data on costs of screening sigmoidoscopy are also available. In a recent, 60-cm flexible sigmoidoscopic study of 412 asymptomatic veterans (mean age 63 years), 22.6 percent had polyps (Gupta et al., 1989). Of these, 63% were adenomatous including two with carcinoma in situ and three with stages A and B carcinoma. It was estimated that the cost of detecting each of these five potentially curable carcinomas was $47,174.

On various grounds, then, recommendations for CRC screening, including adenoma detection, and the expectations for these programs may be unrealistic.

The better targeting of subpopulations at high risk for adenoma formation and CRC has been advocated as one method of rationalizing screening programs.

4. Therapy of Diminutive Polyps

Since it is now recognized that the majority of such polyps (< 0.5 cm in size by definition) are adenomatous rather than hyperplastic, their management is problematic. Although polyps of this size may grow, the chances that they are malignant are very low (only 1 percent of polyps < 1.0 cm show invasive carcinoma). Nevertheless, most workers feel that when such polyps are encountered during colonoscopy, they should be removed (e.g., Waye, 1988a). Spencer et al. (1984) found that subjects harboring small (< 1.0 cm) polyps were not at risk for metachronous CRCs and felt that although the polyps had to be removed, fulguration rather than biopsy was adequate treatment and that the patients required no special surveillance.

In the rectum, relatively more diminutive polyps are hyperplastic than in the colon. It was recently shown (Winawer et al., 1988) that the presence of one or more HyPs in the rectum, without coexisting rectal adenomas, was not accompanied by APs more proximally. The authors concluded that the finding of HyPs during sigmoidoscopy is not an indication for a colonoscopic search for APs. On the other hand Achkar and Carey (1988) found proximal APs on colonoscopy in 29 percent of 72 subjects with HyPs in the rectum but only in 13 percent of 31 patients with essentially normal rectal mucosa. Although the number of subjects examined was small and a p value of but 0.08 between the two groups was recorded, the authors recommended colonoscopy for all individuals in whom sigmoidoscopy revealed small polyps. It is uncertain, however, if the proximal AP prevalence rate (29 percent) they recorded for their hyperplastic polyp group is higher than would be expected in a similarly aged (56 years) control group larger than the one they used. Somewhat more convincing results were provided by Ansher et al. (1989). They found an overall prevelance of APs of 49 percent, and a prevalence of APs proximal to the splenic flexure of 33 percent, in subjects with left-sided hyperplastic polyps; in people without hyperplastic polyps the corresponding prevalences were 15 percent and 3 percent. The issue must thus be regarded as unresolved as of this date.

5. Polypectomy Technique

Electrocautery in one form or another is the method by which the large majority of colorectal polyps that are treated endoscopically are removed. The use of endoscopic laser techniques for the obliteration of adenomas and malignant polyps that are not amenable to electrosurgical removal is at the investigational stage (Mathus-Vliegen & Tytgat, 1986). In all electrosurgical techniques, very high frequency alternating current is delivered to the tissue that is to be resected or destroyed (Odell, 1987). This generates kinetic energy among the intracellular ions, which translates into thermal energy in the affected cells. The resection or ablation of polypoid tissue is the result of thermal damage. Depending on the amount of current and the electrical waveform that is used, electrosurgically treated tissues are first coagulated and then cut. For the purpose of hemostasis, sufficient coagulation is crucial. Premature cutting may cause precipitate resection and serious hemorrhage from uncoagulated blood vessels. The electrical circuit must be completed before current will flow. With monopolar devices (snares and "hot" biopsy forceps), the circuit is completed through the patient to a metal plate, an indifferent electrode that is applied externally, usually to the thigh. In bi- or multipolar devices, the two or more small electrodes are contained within the tip of a probe that is applied to the tissue that is to be coagulated. The circuit is completed through the tissue immediately adjacent to the tip, and no indifferent electrode on the body surface is required. The size and shape of a polyp determine the electrosurgical technique that will be most suitable for its removal.

Snare polypectomy is the method most likely to ensure the complete removal of an adenoma, especially those that are pedunculated. For sessile lesions, a stalk can often be artificially fashioned by pulling the polyp away from the bowel wall, once the snare has been tightened around its base. In either case, the snare is applied to the narrowest neck of tissue, which for a pedunculated adenoma is usually at the top of the stalk, close to the base of the neoplasm; if one attempts to leave a long stalk attached to the polyp by resecting the stalk near the colon wall, a transmural burn may result. Depending on the experience of the endoscopist, it is generally safe to sever stalks with a diameter of up to about 1.5 cm (Williams & Price, 1987; Waye, 1988b) and to ensnare polyps measuring up to 3 cm in greatest dimension. The polyp can then be retrieved in one piece. To achieve this safely, the snare is gradually tightened as current is applied with the electrosurgical unit on a low power setting. In this way, adequate coagulation occurs before the stalk is severed, and hemorrhage is avoided.

Even if severance of the intact polyp is deemed unsafe, piecemeal removal may be possible and is often sufficient treatment, provided the polyp is not malignant. This should not be attempted by the novice colonoscopist.

Diminutive polyps are best managed with the hot-biopsy forceps or bipolar probe. With the former, a polyp is grasped and pulled into the bowel lumen to form a pseudostalk. When current is applied, heating is maximal in the pseudo-stalk, and the polyp is thus severed without damage to the tissue of histological interest, which is retrieved in the cups of the forceps. Once the diagnosis of adenoma has been established, it is probably unnecessary to submit every diminutive polyp from the same person for histological examination; it is common practice to destroy such additional polyps with a bipolar probe.

D. Postpolypectomy Surveillance Recommendations

The main purpose of following polypectomized patients is to prevent the appearance of metachronous carcinomas, especially advanced lesions. The incidence of such carcinomas has been discussed in Chap. 4.A; the recurrence rates and characteristics of new polyps will be discussed here. Removal of new polyps, particularly the larger ones, is desirable in view of their malignant potential. In some studies, only "relevant" recurrent lesions are reported in the follow-up period, that is, the larger (> 0.5 or 1.0 cm) or more dysplastic polyps and the carcinomas (e.g., Matek et al., 1985; Williams & Macrae, 1986).

1. Recurrent Polyps

New polyps are found at a high rate after polypectomy. Annual recurrence rates range from 4 percent to 20 percent (Morson & Bussey, 1985; Wegener et al., 1986; Holtzman et al., 1987; Nava et al., 1987). In a study including a control population the polyp recurrence rate was found to be three times higher than in the controls during the 10-year follow-up period (Brahme et al., 1974). On the average, about 50 percent of American patients will show new polyps in the five years following polypectomy, or about 10 percent per year (Rider et al., 1959; Neugut et al., 1985).

Some authors have found that most recurrences occur early in this period (Waye & Braunfeld, 1982; Neugut et al., 1985; Winawer et al., 1988). It has even been suggested (Neugut et al., 1985) that after the first few years the recurrence rate levels off to approach that of non–polyp-bearing individuals and that surveillance can be slackened at this point. Until more long-term prospective data become available, this does not seem to be advisable.

It is likely that some of the "recurrent" lesions noted in the first few years after polypectomy represent polyps that were not detected during the initial procedure. This possibility stems from the observation of Waye and Braunfeld (1982) that 9 percent of the new polyps discovered within the first year after polypectomy were > 1 cm in size. This suggested to the authors, whose technical expertise is of the highest order, that those lesions were missed at initial colonoscopy, since polyps would not be expected to grow to that size in the space of one year. Most other authors also concede a "miss" rate of about 10 percent. This has implications for surveillance schedules: Many endoscopists recommend that a second colonoscopy be performed within one year of the initial polypectomy to provide confirmation that the colon has been cleared of polyps.

The new polyps are smaller, tend to be more proximal in location, and have a well-differentiated tubular histology (Spjut & Appel, 1979; Wegener et al., 1986; Williams & Macrae, 1986; Winawer et al., 1988). The proximal shift is consistent with autopsy observations showing a more even distribution of polyps than is found in surgical studies and ascribed to the tendency of polyps to arise relatively more frequently in the right colon in later life. One apparently dissenting group (Nava et al., 1987) found that the new polyps occurred in the same colonic segment as the index polyp in 81 percent of the patients, but their study involved only 44 individuals.

2. Risk Factors for New Polyps

Many of the risk factors are the same as for metachronous carcinomas. Multiplicity and large size of the initial polyps are generally accepted factors (Kronborg

TABLE 5.3. Follow-up after colonoscopic polypectomy.

	Time interval (yr)					
	¼	½	1	2	3	5
Pedunculated adenoma					X	
Sessile or pedunculated but questionable complete removal	X				X	
Adenoma with severe dysplasia						
Complete removal		X		X		X
Incomplete or questionably complete removal	X			X		X
Adenoma with invasive cancer completely removed*	X	X	X	X	X	X

*When surgery is not considered.
SOURCE: Modified slightly from Tytgat et al., 1983.

et al., 1983; Lambert et al., 1984; Williams & Macrae, 1986). Villous histology, sessile nature, and severe dysplasia (or carcinoma in the polyp) are more controversial predictors of recurrence. Other factors that have been invoked include old age, male sex, a family history of colorectal neoplasia, and presence of synchronous CRC.

3. Surveillance Schedules

Lifelong colonoscopic surveillance is recommended for all subjects after polypectomy because of the risk of metachronous neoplasms. For subjects at "minimal risk" — that is, with solitary adenomas < 2 cm in size — repeat colonoscopy is recommended after one year to exclude missed lesions. If this is negative, further examinations are performed every three years for six years and every five years thereafter. For patients in the "high-risk" group — that is, those with multiple initial polyps, those with a polyp > 2 cm in diameter, those with sessile villous lesions, or those whose polyps exhibit severe dysplasia or invasive carcinoma (malignant polyps) — repeat colonoscopy is suggested at three to six months and, if negative, at one year after polypectomy. Subsequent colonoscopic surveillance in this group is recommended annually for an additional two years and, if those examinations are negative, every three years thereafter.

A schedule in use in several centers (Frühmorgen & Matek, 1983; Tytgat et al., 1983) is shown in Table 5.3. Others have proposed less complicated schemes, such as colonoscopy every two years for subjects with multiple index adenomas and every four years for those with solitary lesions (Hoff, 1987).

Flexible sigmoidoscopy is not recommended for surveillance purposes because proximal colonic lesions will not be detected, and stool blood tests are too insensitive (Williams & Macrae, 1986).

4. Effectiveness and Cost of Surveillance

The purpose of postpolypectomy colonoscopic surveillance and repeated removal of new APs (and any new carcinomas) is to reduce the incidence of and mortality from CRC in patients harboring polyps. Neither of these objectives has yet been demonstrated. Those prospective colonoscopic studies capable of doing so (e.g., Hermanek et al., 1983; Williams & Macrae, 1986; Winawer et al., 1988) are still in their infancy.

Should colonoscopic surveillance become widespread, a concomitant expansion of facilities and manpower will be required. Assume the prevalence of adenomas is 30 percent among new subjects undergoing total colonoscopy in a typical unit. Assume also that such subjects will have follow-up surveillance colonoscopies every two or four years according to whether the extirpated index lesions were single or multiple. If the same number of new subjects is colonoscoped each year, endoscopic capacity will have to be increased by 90 percent within eight years to accommodate such follow-up examinations.

E. The National Polyp Study

In 1980, a 10-year study was instituted in the United States whose main purpose was to determine optimal surveillance schedules for subjects undergoing polypectomies, the ultimate hope being that repeated colonoscopies with removal of new lesions would reduce incidence of and mortality from CRC (Winawer et al., 1986; Winawer et al., 1988). Another purpose was to gather epidemiological and pathological data on APs and on other polyps, especially HyPs, and to correlate the pathology of the index polyps (risk factors) with the characteristics of new lesions discovered during the follow-up period. Organizations supporting this effort included the American Society for Gastrointestinal Endoscopy, the American Gastroenterological Association, and the American College of Gastroenterology. Multiple institutions were enrolled in the study, representing a variety of health care providers, and included community, veterans and university hospitals, a cancer center, and private practices. Quality assurance centers were established for radiology, pathology, and stool blood testing. Most of the patients were referred for colonoscopy to the participating centers because of symptoms, positive stool blood tests, lesions noted on X ray or endoscopy, or a personal or family history of colorectal neoplasia. Persons with previous polypectomies, a personal history of inflammatory bowel disease, or a personal or family history of FPC and those found to have infiltrating carcinomas or sessile polyps > 3 cm in diameter were excluded from the study. The eligible subjects were stratified according to number (single or multiple) and histological type of adenomas (tubular versus villous). They were then randomized into one group whose follow-up colonoscopy and DCBE were performed both one year and three years later and a second group in whom these procedures were first performed three years later (Figure 5.5).

As of November 1986, 2,342 subjects were included in the data base and 1,001 were randomized. Of 4,484 polyps removed, 68 percent (3,055) were adenomatous. The subjects had a mean age of 62 years; 61 percent were male, and 18 percent had a family history of CRC. Some 43 percent of the patients had multiple

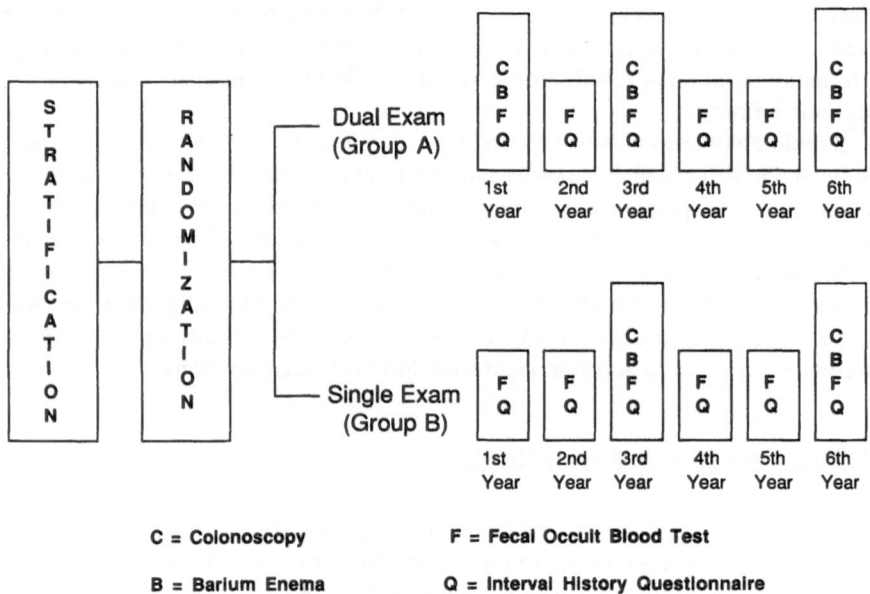

FIGURE 5.5. National Polyp Study. Randomization scheme following stratification according to center, histology, and number of adenomas. (Modified slightly from Winawer et al., 1988.)

APs. Mean adenoma size was 0.9 cm, and 69 percent of the APs were in the left colon (distal to splenic flexure) and 69 percent were tubular. Moderate dysplasia was found in 7 percent, severe dysplasia in 6.9 percent, and infiltrating carcinoma in 1.3 percent of the polyps.

Although the study will last for six years, information at this writing is only available for the first three-year follow-up period. Polyp recurrence rates were 35 percent after three years for the single exam group and 29 percent and 20 percent, respectively, for the first and third years of the dual exam group. The new polyps were smaller, more often tubular, and more proximally located than the initial polyps. Higher recurrence rates were found when the initial polyps were multiple or had villous features or significant dysplasia or when the patients were older or male or had a family history of colorectal neoplasia. Meaningful data on the incidence of carcinoma in the study group must await longer follow-up intervals.

Preliminary data on age and gender distribution, and family histories, of subjects with HyPs indicated striking differences from subjects with APs (discussion in Winawer et al., 1988). Patients with one or more HyPs did not appear to have increased numbers of coincident APs. The only property shared between the two lesions was their distal distribution (initial polyps), but such unrelated tumors as leiomyomas and carcinoids also predominated distally.

F. Polyp Registries

One way to ensure adequate follow-up of subjects with APs is to include them in registries analagous to Tumor Registries. Traditionally, Tumor Registries only include patients with invasive carcinomas, but recently they have accepted subjects with carcinoma in situ, for example, of the uterine cervix, the reasoning being that a substantial number of such individuals will develop invasive carcinoma if they are lost to the medical care system and appropriate treatment. The concept of a registry for a premalignant lesion such as an AP of the colon is an outgrowth of this reasoning and is predicated on the belief that a certain percentage of the patients undergoing polypectomy will develop invasive CRC in the future if they are deprived of follow-up care.

One of the oldest registries is in Erlangen, Germany. This registry was established in 1978 (Hermanek et al., 1983; Hermanek, 1985). All colorectal polyps obtained endoscopically or surgically at the University Hospital of Erlangen were included except those from patients with current or previous CRC or with FPC. At least one follow-up examination was obtained on all of the subjects. Strong emphasis was placed on uniform processing of the polyps, including adequate numbers of step sections to provide representative histological sampling of the specimens as described in Chap. 5.B. Multivariate analysis of the results was carried out using logit analysis in order to quantify the interactions between the various risk factors in malignant polyps. Types of polyps found are shown in Table 2.2.

It was found that carcinoma was found more frequently in APs with large size, sessile configuration, villous histology, and rectal location. None out of 2,214 APs < 5 mm had carcinoma, and such polyps were thus considered devoid of clinical significance. They found by multivariate analysis that rectal location alone did not favor the occurrence of carcinoma in the polyps and that the increased carcinoma rate in that location could be explained by the fact that polyps there were larger and more villous than in more proximal colonic segments. They did not analyze separately for degree of dysplasia since they felt, again, that dysplasia correlated with other known risk factors such as size and villosity and that its assessment had a considerable subjective component. (I would certainly agree with the latter statement.)

Another finding of interest was that multiplicity of polyps, considered by most to be a risk factor for current and metachronous carcinomas, was actually associated with a *lower*-than-expected probability of carcinoma in a given adenoma. Multiplicity is still a valid risk factor for the *patient*, however. In fact, their follow-up studies indicated that subjects with multiple adenomas tended to develop "relevant" findings (i.e., polyps > 5 mm or with severe dysplasia, or carcinomas) sooner than subjects with solitary adenomas. This suggested to those workers that two-year postpolypectomy follow-up examination should be performed for subjects with multiple polyps, whereas four-year follow-up was adequate for those with solitary polyps.

Another Polyp Registry has been established at St. Mark's Hospital in London (Williams & Macrae, 1986). Intially, 500 patients with APs were included in a *pilot study* whose purposes included investigation of the optimal postpolypectomy follow-up interval (one, three, or five years), evaluation of various diagnostic techniques, and the establishment of risk factors allowing inclusion of subjects into high- and low-risk groups. Since the study was not randomized and did not include a control group, it was difficult to determine the incidence of new polyps and carcinomas. Results indicated that (1) stool occult blood tests, history taking (primarily for bleeding), and rigid proctosigmoidoscopy were relatively ineffective in detecting neoplasias; (2) the major risk factors of use in predicting new adenomas > 1 cm or carcinomas ("significant lesions") included male sex, advanced age, and multiple or large initial polyps but not severe dysplasia or villous component; and (3) total colonoscopy was the best means of detecting new lesions, although a combination of flexible sigmoidoscopy and DCBE were useful in the 30 percent of subjects in whom technical problems were encountered with colonoscopy.

The same group has established a *long-term* study of 800 patients in which the methodology and goals have been somewhat modified because of the pilot study results. Because of missed initial lesions, each subject will receive two colonoscopies, or one colonoscopy and a DCBE, within an 18-month period to ensure an adenoma-free colon before entry into the study. All colonoscopies will be performed by C.B. Williams. Risk factors will be reexamined to determine the validity of the pilot study results. Total colonoscopy will be the preferred follow-up method; flexible sigmoidoscopy plus DCBE will only be used when there are technical problems with colonoscopy. Subjects will be assigned to a high-risk group—that is, those (1) over 60 years in age, (2) over 55 years with two or more previous adenomas, and (3) under 55 years with five or more adenomas—and a low-risk group—which will include all others. High-risk subjects will be examined at one- and three-year intervals, and low-risk subjects at three- or five-year follow-up intervals. An age- and sex-matched group of adenoma-free controls will be included and compared with the low-risk group to determine whether the latter require any special surveillance at all.

Preliminary results (C.B. Williams, personal communication, 1988) indicate that the overall polyp recurrence rate is 30 percent, the average adenoma size is 0.4 cm at five years, and 97 percent of the new adenomas are < 1 cm. One adenoma, presumably a lesion missed initially, was 1.2 cm, and two carcinomas measuring 0.7 and 1.0 cm in diameter have been detected. The tentative impression of those investigators is that the recurrence rate for patients with solitary adenomas is so low that any follow-up at all may not be justified. A subset of individuals of "superhigh risk" with > 10 adenomas is being investigated to determine the need for more frequent examinations in that group.

In 1987, a Polyp Registry was established within the framework of the Tumor Registry at the Roger Williams General Hospital in Rhode Island. Its purpose was to enroll patients whose adenomatous polyps were submitted to the Pathology Department of the hospital. The patients are to be followed in a manner similar

TABLE 5.4. Histological characteristics of 191 adenomatous polyps from 105 subjects enrolled in 1984 in the Rhode Island Polyp Registry.

	% of total number of polyps[a]
Degree of dysplasia	
Mild	64.4
Moderate	25.1
Severe[b]	9.9
Histologic type	
Tubular	69.6
Tubulovillous	19.4
Villous	11.0

[a] Figures rounded off to nearest decimal point.
[b] Plus one malignant polyp.

to those in the Tumor Registry in order to ensure that new polyps and (early) carcinomas will be detected and removed. The ultimate goal is to reduce mortality from CRC in the enrollees, although we realize that it will require large numbers of subjects and prolonged follow-up to demonstrate this. We also wish to obtain epidemiological and pathological information on the initial polyps and recurrent lesions and hope to detect changing modes of surveillance with time (e.g., a shift from rigid to flexible sigmoidoscopes and from sigmoidoscopy to colonoscopy). The educational components will include a narrative and pictorial representation on dietary recommendations for patients, and recommended postpolypectomy surveillance management and schedules for the participating physicians. The schedules, which will be revised periodically as new information becomes available, will be derived from guidelines published by the American Society for Gastrointestinal Endoscopy, other major Gastroenterology Societies, and groups such as the American Cancer Society and the National Cancer Institute.

Subjects to be admitted to the Polyp Registry will first be examined for eligibility. Those with current or previous CRC, with FPC, or with inflammatory bowel disease will be excluded. Demographic, historical, and pathological data will be entered into our IBM computer using the same software program (Rocky Mt. Cancer Data System) used in the Tumor Registry. The initial cohort includes all eligible patients who had adenomatous polyps removed in 1984 and who will be followed initially for three years. That group comprises 105 subjects with 191 adenomatous polyps. The mean age of the group is 65 years, and it includes 67 men and 38 women. Of these subjects, 40 percent have more than one polyp, and 44 percent have polyps > 1 cm in size. Other data are shown in Table 5.4. Preliminary four-year follow-up data are available on 98 of the subjects. Eighteen patients have died, none with CRC. One live patient developed CRC at a site different from that of the original polyp. Of the remaining live subjects, one or more recurrent APs were found in 37.8 percent; 65.3 percent had one or more colonoscopies performed, 29.6 percent had one or more sigmoidoscopic examinations, and 31.6 percent had one or more stool blood tests.

References

Achkar E, Carey W (1988) Small polyps found during fiberoptic sigmoidoscopy in asymptomatic patients. Ann Intern Med 109:880–883.

Ahlquist DA, McGill DB, Schwartz S, Taylor WF, Owen RA (1985) Fecal blood levels in health and disease: A study using HemoQuant. N Engl J Med 312:1422–1428.

American Cancer Society (1980) Guidelines for the cancer-related checkup. Recommendations and rationale. CA 30:194–240.

American Society for Gastrointestinal Endoscopy Guidelines (1986) The role of colonoscopy in the management of patients with colonic polyps.

Ansher AF, Lewis JH, Fleischer DE, Cattau EL, Collen MJ, O'Kieffe DA, Korman LY, Benjamin SB (1989) Hyperplastic colonic polyps as a marker for adenomatous colonic polyps. Am J Gastroenterol 84:113–117.

Applegate WB, Spector MH (1981) Colorectal cancer screening. J Community Health 7:138–151.

Barry MJ, Mulley AG, Richter JM (1987) Effect of work up strategy on the cost-effectiveness of fecal occult blood screening for colorectal cancer. Gastroenterology 93:301–310.

Bartram CI, Kumar P (1981) Clinical Radiology in Gastroenterology, 1st Ed. Blackwell Scientific Publications, Oxford.

Befeler D (1967) Proctoscopic perforation of the large bowel. Dis Colon Rectum 10:376–378.

Bertario L, Spinelli P, Gennari L, Sala P, Pizzetti P, Severini A, Cozzi G, Bellomi M, Berrino F (1988) Sensitivity of Hemoccult test for large bowel cancer in high-risk subjects. Dig Dis Sci 33:609–613.

Brahme F, Ekelund GR, Norden JG, Wenckert AM (1974) Metachronous colorectal polyps: Comparison of development of colorectal polyps and carcinomas in persons with and without histories of polyps. Dis Colon Rectum 17:166–171.

Bremond A, Collet P, Lambert R, Martin JL (1984) Breast cancer and polyps of the colon. Cancer 54:2568–2570.

Burt RW, Bishop DT, Cannon LA, Dowdle MA, Lee RG, Skolnick MH (1985) Dominant inheritance of adenomatous colonic polyps and colorectal cancer. N Engl J Med 312:1540–1544.

Cannon-Albright LA, Skolnick MH, Bishop DT, Lee RG, Burt RW (1988) Common inheritance of susceptibility to colonic adenomatous polyps and associated colorectal cancers. N Engl J Med 319:533–537.

Carroll R, Klein M (1980) How often should patients be sigmoidoscoped? A mathematical perspective. Prev Med 9:741–746.

Chait M, Herbert E, Guthrie M, Flehinger B, Winawer SJ (1986) The Memorial Sloan-Kettering Cancer Center-Strang Clinic Program: A progress report. Front Gastrointest Res 10:102–111.

Chamberlain J, Day NE, Hakama M, Miller AB, Prorok PC (1986) UICC workshop of the project on evaluation of screening programmes for gastrointestinal cancer. Int J Cancer 37:329–334.

Christie JP (1984) Malignant colon polyps—cure by colonoscopy or colectomy? Am J Gastroenterol 79:543–547.

Christie JP (1988) Polypectomy or colectomy? Management of 106 consecutively encountered malignant colorectal polyps. Am Surg, Feb:93–99.

Colacchio TA, Forde KA, Scantlebury VP (1981) Endoscopic polypectomy; inadequate treatment for invasive colorectal carcinoma. Ann Surg 194:704–707.

Cooper HS (1983) Surgical pathology of endoscopically removed malignant polyps of the colon and rectum. Am J Surg Pathol 7:613–623.

Coustifides T, Lavery I, Benjamin SP, Sivak MV (1979) Malignant polyps of the colon and rectum: A clinicopathology study. Dis Colon Rectum 22:82–86.

Cranley JP, Petras RE, Carey WD, Paradis K, Sivak MV (1986) When is endoscopic polypectomy adequate therapy for colonic polyps containing invasive carcinoma? Gastroenterology 91:419–427.

Dales LG, Friedman GD, Collen MF (1979) Evaluating periodic multiphasic health checkups: A controlled trial. J Chronic Dis 32:385–404.

Day D, Morson B (1978) The pathology of adenomas. In B Morson (ed): The Pathogenesis of Colorectal Cancer. WB Saunders, Philadelphia, pp 43–57.

DeCosse JJ (1984) Malignant colorectal polyps. Gut 25:433–435.

DeCosse JJ (1988) Early cancer detection. Cancer 62:1787–1790.

Eddy DM (1984) Cost-effectiveness of colorectal cancer screening. In B Levin, RH Riddell (eds): Frontiers in Gastrointestinal Cancer. Elsevier, New York.

Farrands PA, Vellacott KD, Amar SS, Balfour TW, Hardcastle JD (1983) Flexible fiberoptic sigmoidoscopy and double-contrast barium-enema examination in the identification of adenomas and carcinoma of the colon. Dis Colon Rectum 26:727–729.

Flehinger BJ, Herbert E, Winawer SJ, Miller DG (1988) Screening for colorectal cancer with fecal occult blood test and sigmoidoscopy: Preliminary report of the colon project of Memorial Sloan Kettering Cancer Center and PMI-Strang Clinic. In J Chamberlain, AB Miller (eds): Screening for Gastrointestinal Cancer. Hans Huber, Toronto, pp 9–16.

Fleisher M, Schwartz MK, Winawer SJ (1985) Fecal occult blood testing. In AB Miller (ed): Screening for Cancer. Academic Press, New York, pp 237–247.

Fork F (1983) Reliability of routine double contrast examination of the large bowel: A prospective study of 2590 patients. Gut 24:672–677.

Frank JW (1985) Occult-blood screening for colorectal carcinoma: The yield and the costs. Am J Prev Med 1:18–24.

Friedman GD, Collen MF, Fireman BH (1986) Multiphasic health checkup evaluation: A 16-year follow-up. J Chronic Dis 39:453–463.

Frühmorgen P, Matek W (1983) Significance of polypectomy in the large bowel—endoscopy. Endoscopy 15:155–157.

Gabrielson N, Granqvist S, Nilsson B (1985) Guaiac test for detection of occult faecal blood loss in patients with endoscopically verified colonic polyps. Scand J Gastroenterol 20:978–982.

Gilbertsen VA (1979) The earlier detection of colorectal cancers. In DR Brodie (ed): Screening and Early Detection of Colorectal Cancer. US Department of Health, Education and Welfare, National Cancer Institute, Washington, DC, pp 211–215.

Gilbertsen VA, McHugh RB, Schuman LM, Williams SE (1980) Colon cancer control study: An interim report. In S Winawer, D Schottenfeld, P Sherlock (eds): Colorectal Cancer: Prevention, Epidemiology and Screening. Raven Press, New York, pp 261–270.

Gilbertsen VA, Nelms JM (1978) The prevention of invasive cancer of the rectum. Cancer 41:1137–1139.

Greenwald P, Sondik EJ (eds) (1986) Cancer control objectives for the nation: 1985–2000. III. Screening. Natl Cancer Inst Monogr 2:27–32.

Griffin JW (1985) Flexible fiberoptic sigmoidoscopy— longer may not be better for the nonendoscopist. Gastrointest Endosc 31:347–348.

Grinnell RS, Lane N (1958) Benign and malignant adenomatous polyps and papillary adenomas of the colon and rectum. An analysis of 1,856 tumors in 1,335 patients. Int Abst Surg 106:519–538.

Grossman S, Milos ML (1988) Colonoscopic screening of persons with suspected risk factors for colon cancer. Gastroenterology 94:395–400.

Gupta TP, Jaszewski R, Luk GD (1989) Efficacy of screening flexible sigmoidoscopy for colorectal neoplasia in asymptomatic subjects. Am J Med 86:547–550.

Haggitt RC, Glotzbach RE, Soffer EE, Wruble LD (1985) Prognostic factors in colorectal carcinomas arising in adenomas: Implications for lesions removed by endoscopic polypectomy. Gastroenterology 89:328–336.

Hallstrom AP, Shaneyfelt SL, Mahler AK, Silverstein FE (1984) The national ASGE colonoscopy survey—complications of colonoscopy. Gastrointest Endosc 30:156 (Abstract).

Hardcastle JD, Armitage NC, Chamberlain J, Amar SS, James PD, Balfour TW (1986) Fecal occult blood screening for colorectal cancer in the general population. Cancer 58:397–403.

Heberer G, Wiebecke B, Zumtobel V, Hamperl D (1978) Maligne Polypen und früher-fasste Karzinome des Rektums. Münch Med Wschr 120:201–206.

Hermanek P (1985) Diagnosis and therapy of cancerous adenoma of the large bowel: A German experience. In CM Fenoglio-Preiser, FP Rossini (eds): Adenomas and Ade-nomas Containing Carcinoma of the Large Bowel: Advances in Diagnosis and Therapy. Raven Press, New York, pp 57–62.

Hermanek P, Frühmorgen P, Guggenmoos-Holzmann I, Altendorf A, Matek M (1983) The malignant potential of colorectal polyps—a new statistical approach. Endoscopy 15:16–20.

Herzog P, Holtermuller KH, Preiss J, Fischer J, Ewe K, Schreiber HJ, Berres M (1982) Fecal blood in patients with colonic polyps: A comparison of measurements with ^{51}chromium-labeled erythrocytes and with the haemoccult test. Gastroenterology 83:957–962.

Hoff G (1987) Colorectal polyps. Clinical implications: Screening and cancer prevention. Scand J Gastroenterol 22:769–775.

Holtzman R, Poulard JB, Bank S, Levin LR, Flint GW, Strauss RJ, Margolis IB (1987) Repeat colonoscopy after endoscopic polypectomy. Dis Colon Rectum 30:185–188.

Isbister WH (1986) Colorectal polyps: An endoscopic experience. Aust NZ J Surg 56:717–722.

Katon RM (1979) Sigmoidoscopy (rigid and flexible). In DR Brodie (ed): Screening and Early Detection of Colorectal Cancer. US Department of Health, Education and Welfare, National Cancer Institute, Washington, DC, pp 115–123.

Knight KK, Fielding JE, Battista RN (1989) Occult blood screening for colorectal cancer. JAMA 261:587–593.

Knutson CO, Max MH (1979) Diagnostic and therapeutic colonoscopy. A critical review of 662 examinations. Arch Surg 114:430–435.

Kronborg O, Hage E, Adamsen S, Deichgraeber E (1983) Follow-up after colorectal poly-pectomy. Scand J Gastroenterol 18:1095–1099.

Lambert R, Sobin L, Waye J, Stalder G (1984) The management of patients with colorectal adenomas. Cancer 34:167–176.

Lehman GA, Buchner DM, Lappas JC (1983) Anatomical extent of fiberoptic sigmoidoscopy. Gastroenterology 84:803–808.

Lipper S, Kahn LB, Ackerman LV (1983) The significance of microscopic invasive cancer in endoscopically removed polyps of the large bowel: A clinicopathologic study of 51 cases. Cancer 52:1691–1699.

Macrae FA, St. John DJB (1982) Relationship between patterns of bleeding and hemoccult sensitivity in patients with colorectal cancers or adenomas. Gastroenterology 82:891–898.

Macrae F, Tan K, Williams C (1983) Towards safer colonoscopy: A report on the complications of 5000 diagnostic or therapeutic colonoscopies. Gut 24:376–383.

Macrae FA, Williams CB (1985) Sigmoidoscopy and other tests for colorectal cancer. In AB Miller (ed): Screening for Cancer. Academic Press, New York, pp 249–269.

Marshak RH (1981) More on diminutive colonic polyps. Gastroenterology 80:628.

Matek W, Guggenmoos-Holzmann I, Demling L (1985) Follow-up of patients with colorectal adenomas. Endoscopy 17:175–181.

Mathus-Vliegen EMH, Tytgat GNJ (1986) Nd:YAG laser photocoagulation in colorectal adenoma. Evaluation of its safety, usefulness and efficacy. Gastroenterology 90: 1865–1873.

Miller AB (1988) Review of sigmoidoscopic screening for colorectal cancer. In J Chamberlain, AB Miller (eds): Screening for Gastrointestinal Cancer. Hans Huber, Toronto, pp 3–7.

Minsky BD, Mies C, Recht A, Rich TA, Chaffey JT (1988) Resectable adenocarcinoma of the rectosigmoid and rectum. Cancer 61:1417–1424.

Morrison AS (1985) Screening in Chronic Disease. Oxford University Press, New York.

Morson BC, Bussey HJR (1985) Magnitude of risk for cancer in patients with colorectal adenomas. Br J Surg [Suppl] 72:23–28.

Morson BC, Bussey HJR, Day DW, Hill MJ (1983) Adenomas of large bowel. Cancer Surv 2:451–478.

Morson BC, Bussey HJR, Samoorian S (1977) Policy of local excision for early cancer of the colorectum. Gut 18:1045–1050.

Morson BC, Whiteway JE, Jones EA, Macrae FA, Williams CB (1984) Histopathology and prognosis of malignant colorectal polyps treated by endoscopic polypectomy. Gut 25:437–444.

Muto T, Bussey HJR, Morson BC (1975) The evolution of cancer of the colon and rectum. Cancer 36:2251–2270.

Muto T, Kamiya J, Sawada T, Kusama S, Itai Y, Ikenaga T, Yamashiro M, Hino Y, Yamaguchi S (1980) Colonoscopic polypectomy in diagnosis and treatment of early carcinoma of the large intestine. Dis Colon Rectum 23:68–75.

Nava H, Carlson G, Petrelli NJ, Herrera L, Mittelman A (1987) Follow-up colonoscopy in patients with colorectal adenomatous polyps. Dis Colon Rectum 30:465–468.

Nelson RN, Abcarian H, Prasad ML (1982) Iatrogenic perforation of the colon and rectum. Dis Colon Rectum 25:305–308.

Neugut A, Johnsen C, Forde KA, Treat MR (1985) Recurrence rates for colorectal polyps. Cancer 55:1586–1589.

Neugut AI, Pita S (1988) Role of sigmoidoscopy in screening for colorectal cancer: A critical review. Gastroenterology 95:492–499.

Odell RC (1987) Principles of electrosurgery. In MV Sivak (ed): Gastroenterologic Endoscopy, 21st Ed. WB Saunders, Philadelphia.

Olson RM, Perencevich NP, Malcom AW, Chaffey JT, Wilson RE (1980) Patterns of recurrence following curative resection of adenocarcinoma of the colon and rectum. Cancer 45:2969–2974.

Richards WO, Webb WA, Morris SJ, Davis RC, McDaniel L, Jones L, Littauer S (1987) Patient management after endoscopic removal of the cancerous colon adenoma. Ann Surg 205:665–670.

Rickert RR, Auerbach O, Garfinkel L, Hammond EC, Frasca JM (1979) Adenomatous lesions of the large bowel. An autopsy survey. Cancer 43:1847–1857.

Riddell RH (1985) Hands off "cancerous" large bowel polyps. Gastroenterology 89:432–435.

Rider JA, Kirsner JB, Moeller JC, Palmer WL (1959) Polyps of the colon and rectum. A four-year to nine-year follow-up study of five hundred thirty seven patients. JAMA 170:633–638.

Sanowski RA, Groveman HS, Klauber M (1985) Training primary care physicians in flexible sigmoidoscopy. Gastrointest Endosc 31:149. (Abstract)

Schottenfeld D, Winawer SJ (1982) Large intestine. In D Schottenfeld, JR Fraumeni, Jr (eds): Cancer Epidemiology and Prevention. Saunders, Philadelphia, pp 703–727.

Selby JV, Friedman GD (1989) Sigmoidoscopy in the periodic health examination of asymptomatic adults. JAMA 261:595–601.

Shinya H, Wolff WI (1979) Morphology, anatomic distribution, and cancer potential of colonic polyps. Ann Surg 190:679–683.

Simon JB (1985) Occult blood screening for colorectal carcinoma: A critical review. Gastroenterology 88:820–837

Sobin LH (1985) The histopathology of bleeding from polyps and carcinomas of the large intestine. Cancer 55:577–581.

Spencer RJ, Melton LJ, Ready RL, Ilstrup DM (1984) Treatment of small colorectal polyps: A population-based study of the risk of subsequent carcinoma. Mayo Clin Proc 59:305–310.

Spjut HJ, Appel M (1979) Epithelial polyps of the large bowel: A pathological and colonoscopic study. Current Probl Cancer 4:23–42.

Spjut HJ, Estrada RG (1977) The significance of epithelial polyps of the large bowel. Pathol Annu, Pt I, 12:147–170.

Steele G, Osteen RT (1986) Surgical treatment of colon cancer. In G Steele, RT Osteen (eds): Colorectal Cancer: Current Concepts in Diagnosis and Treatment. Marcel Dekker, New York, pp 127–162.

Talbot IC, Ritchie S, Leighton M, Hughes AO, Bussey HJR, Morson BC (1981) Invasion of veins by carcinoma of rectum: Method of detection, histological features and significance. Histopathology 5:141–163.

Tytgat GNJ, Mathus-Vliegen EMH, Offerhaus J (1983) Value of endoscopy in the surveillance of high-risk groups for gastrointestinal cancer. In P Sherlock, BC Morson, L Barbara, U Veronesi (eds): Precancerous Lesions of the Gastrointestinal Tract. Raven Press, New York, pp 305–318.

Waye JD (1988a) Management of the diminutive polyp. Proceedings of the 3rd International Congress on Colonoscopy and Disease of the Large Bowel, March 3–5, 1988, Bal Harbour, FL.

Waye JD (1988b) Techniques of colonoscopy, hot biopsy forceps, and snare polypectomy. In G Steele, RW Burt, SJ Winawer, JP Karr (eds): Basic and Clinical Perspectives of Colorectal Polyps and Cancer. Alan R Liss, New York, pp 61–69.

Waye JD, Braunfeld S (1982) Surveillance intervals after colonoscopic polypectomy. Endoscopy 14:79–81.

Wegener M, Borsch G, Schmidt G (1986) Colorectal adenomas: Distribution, incidence of malignant transformation, and rate of recurrence. Dis Colon Rectum 29:383–387.

Wilcox GM, Anderson PA, Colacchio TA (1986) Early invasive carcinoma in colonic polyps. A review of the literature with emphasis on the assessment of the risk of metastasis. Cancer 57:160–171.

Wilcox GM, Beck JR (1987) Early invasive cancer in adenomatous colonic polyps ("malignant polyps"). Evaluation of the therapeutic options by decision analysis. Gastroenterology 92:1159–1168.

Williams CB, Macrae FA (1986) The St. Mark's neoplastic polyp follow-up study. Front Gastrointest Res 10:226–242.

Williams CB, Macrae FA, Bertram C (1982) A prospective study of diagnostic methods in adenoma follow-up. Endoscopy 14:74–78.

Williams CB, Price AB (1987) Colon polyps and carcinoma. In MV Sivak (ed): Gastroenterologic Endoscopy, 1st Ed. WB Saunders, Philadelphia.

Williams CB, Whiteway JE (1985) Conservative management of malignant polyps: St. Mark's results at 5 years. In CM Fenoglio-Preiser, FP Rossini (eds): Adenomas and Adenomas Containing Carcinoma of the Large Bowel: Advances in Diagnosis and Therapy. Raven Press, New York, pp 67–73.

Wilson RE (1986) Prognostic factors and natural history of colorectal cancer. In G Steele, RT Osteen (eds): Colorectal Cancer. Current Concepts in Diagnosis and Treatment. Marcel Dekker, New York, pp 99–125.

Winawer SJ (1979) Progress report of controlled trial of screening with fecal occult blood testing. In DR Brodie (ed): Screening and Early Detection of Colorectal Cancer. US Department of Health, Education and Welfare, National Cancer Institute, Washington, DC, pp 193–210.

Winawer SJ, Andrews M, Miller CH, Fleisher M (1980) Review of screening for colorectal cancer using fecal occult blood testing. In D Schottenfeld, P Sherlock (eds): Colorectal Cancer: Prevention, Epidemiology and Screening. Raven Press, New York, pp 249–259.

Winawer SJ, Ritchie MT, Diaz BJ, Gottlieb LS, Stewart ET, Zauber A, Herbert E, Bond J (1986) The National Polyp Study: Aims and organization. Front Gastrointest Res 10:216–225.

Winawer SJ, Zauber A, Diaz B, O'Brien M, Gottlieb LS, Sternberg SS, Waye JD, Shike M, National Polyp Study Work Group (1988) The National Polyp Study: Overview of program and preliminary report of patient polyp characteristics. In G Steele, RW Burt, SJ Winawer, JP Karr (eds): Basic and Clinical Perspectives of Colorectal Polyps and Cancer. Alan R Liss, New York, pp 23–33.

Winman G, Berci A, Panish J, Talbot TM, Overholt BF, McCallum RW (1980) Superiority of the flexible to the rigid sigmoidoscope in routine proctosigmoidoscopy. N Engl J Med 302:1111–1112.

6
Conclusions

Science . . . unlike poetry . . . does not seek to exploit its ambiguities, but to minimize them. (Bronowski, 1965, p. 49)

The main conclusion of this book is that adenomatous epithelium represents the most common precursor of colorectal cancer. Small (< 0.5 cm) or microscopic carcinomas devoid of adenomatous tissue and designated de novo lesions are very rare indeed in comparison with similar-sized malignant foci occurring in pre-existing adenomatous polyps. Most commonly, adenomatous tissue presents as a polyp, but exceptions are found. In hereditary nonpolyposis colorectal cancer, "flat adenomas" have been described, and in ulcerative colitis some of the dys-plastic lesions found in the flat mucosa consist of adenomatous epithelium indis-tinguishable from that found in APs.

The classic AP can thus be viewed as a grossly visible marker for risk of CRC. The existence of such a marker should be viewed as a fortunate occurrence since detectable premalignant lesions are not present for most malignancies, for exam-ple, carcinomas of the lung or pancreas. The subject with such a polyp is at risk first for carcinoma developing in the polyp. The majority of APs, however, are not premalignant, and many do not appear to grow at all. For the minority that show malignant transformation, the adenoma–carcinoma interval has been esti-mated at around 10 years, which is similar to that of other established premalig-nant lesions such as carcinoma in situ of the uterine cervix. Since most polyps are removable endoscopically, however, there is no need to postpone their excision and monitor them radiographically. This had been the policy for many polyps in the precolonoscopic era when the morbidity and mortality of laparotomy often exceeded the risk of malignant change.

The magnitude of the risk of carcinomatous transformation will depend on the number of risk factors (e.g., multiplicity, large size, villosity, severe dysplasia, and so on). The presence of these factors will also determine the likelihood of the second major risk borne by polyp-bearers, namely metachronous carcinoma else-where in the colon. Evaluation of certain risk factors, however, such as degree of villosity and severity of dysplasia, is quite subjective and shows considerable interobserver variation despite the presence of criteria described in the World

Health Organization guidelines and elsewhere. Although an attempt has been made to quantify the villous component in APs, this assessment is dependent on one's definition of what constitutes villi and on adequacy of polyp sampling, which is especially critical for larger polyps.

Numerous studies have documented biochemical, histochemical, ultrastructural, and kinetic changes in APs and in other premalignant dysplasias (e.g., in ulcerative colitis) and in normal mucosae of subjects with colorectal neoplasia. It has been suggested that some of these biomarkers may help focus screening and surveillance efforts aimed at detection of future CRC in such subjects, but this benefit has not yet been demonstrated in any prospective study to date. Moreover, many of the changes in APs and in dysplastic foci in ulcerative colitis do not contribute information beyond that already revealed by routine histological examination.

Many investigators, including the author of this volume, feel that since the majority of CRCs originate in APs, prophylactic polypectomy with repeated polypectomies as new polyps appear should theoretically at least reduce the incidence of CRC. It would thus seem profitable to identify subjects at risk for developing APs (as well as CRCs). Such groups will include those with a personal history of colorectal neoplasia or a family history of familial polyposis coli and those > 40 years old. (There is controversy on the degree of adenoma risk in subjects with a family history of HNCC or of sporadic polyps and CRCs, many of which are now also believed to be genetically determined.) The financial costs and manpower requirements required to detect and remove polyps in such populations may be considerable, however.

The current screening techniques used to detect adenomas in those groups include the fecal occult blood test (FOBT) and flexible sigmoidoscopy. The Hemoccult test is an insensitive method of detecting polyps, especially small ones, but newer stool blood tests may prove to be more successful in doing so. These techniques when followed by postpolypectomy surveillance with colonoscopy, or combined flexible sigmoidoscopy and barium enema, are most likely to succeed in maintaining the colon free of adenomas. Unfortunately, neither of these screening modalities (FOBT and sigmoidoscopy), even when accompanied by repeated screenings and removal of new polyps, has as yet met the criterion of lowering CRC mortality in the screened populations. Although the detection of large (potentially malignant) polyps and early-stage carcinomas has been described in these studies, most investigators do not feel that these findings justify mass population screening programs. However, some of these screening studies have not yet been completed. It is possible that their final results and those of the National Polyp Study in the United States, and data from the Polyp Registries in various countries, may provide new information on the value of screening and surveillance.

Reference

Bronowski J (1965) The Identity of Man. Penguin Books, Middlesex, England.

Glossary and Abbreviations

Adenomatous polyp (AP) (or adenoma). Comprises tubular adenoma, tubulovillous adenoma, and villous adenoma.

Barium enema (BE)

Cancer family syndrome (CFS). Now also called Lynch syndrome II, a variant of HNCC.

Carcinoembryonic antigen (CEA). An oncofetal antigen found in serum of some patients with CRC and frequently detectable histochemically in adenomatous and carcinomatous tissue.

Colorectal cancer (or carcinoma) (CRC)

de novo carcinoma. CRC arising in flat, nonpolypoid mucosa.

Double contrast barium enema (DCBE). Preferred over single contrast BE for detection of colorectal polyps.

Familial polyposis coli (FPC). Autosomal dominant condition characterized by numerous (>100) APs of colon.

Fecal occult blood test (FOBT). Screening test for colorectal neoplasia.

Gardner's syndrome (GS). Variant of FPC with extracolonic manifestations.

Hereditary nonpolyposis colon cancer (HNCC). Subdivided into Lynch syndrome I (no extracolonic cancers) and Lynch syndrome II (accompanied by a variety of extracolonic cancers).

Hyperplastic polyp (HyP). Common non-neoplastic polyp found alone or in subjects with CRC.

Inflammatory bowel disease. Comprises ulcerative colitis and Crohn's disease. Patients with inflammatory bowel disease are prone to develop CRC.

Metachronous APs or CRCs. The occurrence at different times and different sites of two neoplastic lesions (i.e., not recurrences).

Predictive value (PV). The proportion of people with a positive screening test who have the disease in question.

Restriction–fragment–length polymorphisms (RFLP).   A chromosome mapping technique that has been used to study colonic neoplasia-prone populations.

Sporadic AP or CRC.   Also called "common" polyp or carcinoma. The occurrence of a neoplastic lesion in the absence of a documented familial history of HNCC, FPC, or allied disorders or in the absence of other predisposing factors such as inflammatory bowel disease.

Synchronous APs or CRCs.   The simultaneous occurrence of two neoplastic lesions.

TPA.   12-0-tetracanoylphorbol-13-acetate. A phorbol ester used as a tumor promoter in experimental carcinogenesis.

Transitional mucosa.   The mucosa bordering CRCs and APs. Some workers believe that specific premalignant changes can be found in this mucosa, but there is considerable opposition to this theory.

Tubular adenoma (TA).   An adenomatous polyp consisting primarily of tubules (e.g., > 75 percent).

Tubulovillous adenoma (TVA).   An adenomatous polyp with substantial numbers of tubules and villi (e.g., from 25 to 75 percent tubules).

Ulcerative colitis (UC)

Villous adenoma (VA).   An adenomatous polyp consisting primarily of villi (e.g., > 75 percent).

Index